Decoding 'The God'

and

'The Religion'

(Volume-1)

AKHIL RAJENDRA

Leadstart
INKSTATE

ISBN 978-93-90266-76-0
Copyright © Akhil Rajendra, 2020

First published in India 2021 by Leadstart Inkstate
A Division of One Point Six Technologies Pvt Ltd

Sales Office:
Unit No.25/26, Building No.A/1,
Near Wadala RTO,
Wadala (East), Mumbai – 400037 India
Phone: +91 969933000
Email: info@leadstartcorp.com
www.leadstartcorp.com

Disclaimer: The views expressed in this book are those of the Author and do not pertain to be held by the Publisher.

Cover: Nitin Ingale
Layouts: Victor Patali

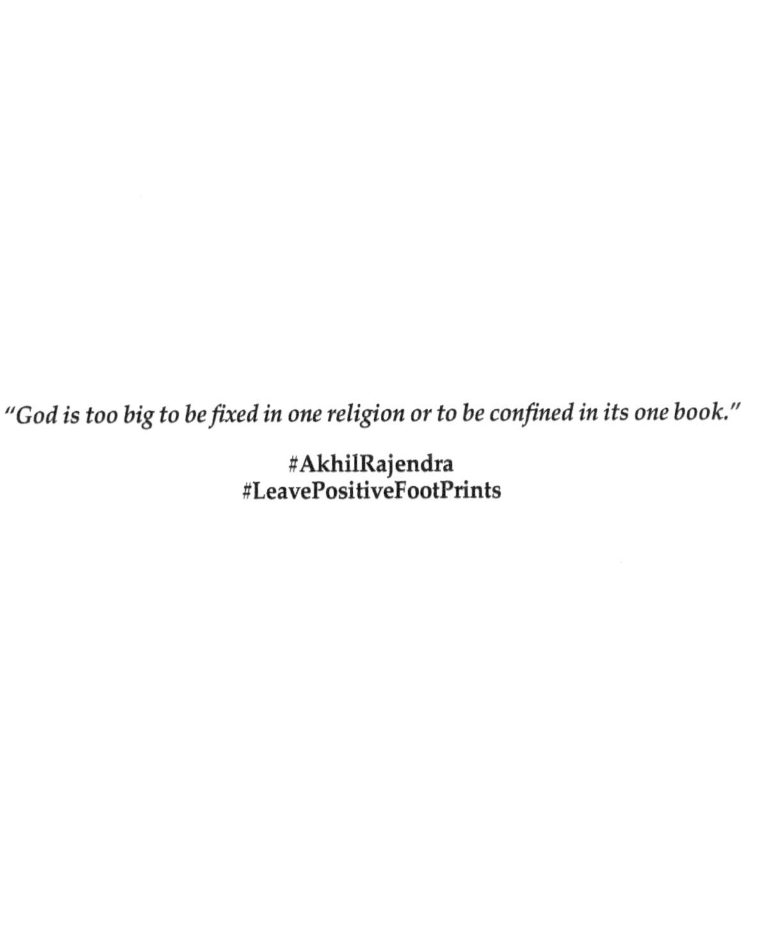

"God is too big to be fixed in one religion or to be confined in its one book."

#AkhilRajendra
#LeavePositiveFootPrints

ABOUT THE AUTHOR

Akhil Rajendra, based at Mumbai, is the author of book series Decoding 'The God' and 'The Religion', He is a science graduate and post graduate in business management, chose creative field to work and explore professionally, let that be advertising in various mediums, films, theatre, screen writing or book writing, he always worked where his passion guided him.

Many people from various grounds and professions, including of country's most eminent personalities have been taking astrological, 'spiritual advices and healings' from Akhil for many years, but till date he never considered his knowledge and skills as a profession and never took any fee or any other materialistic favour for such helps and guidance.

Decoding 'The God' and 'The Religion' series is very close to Akhil's heart, lockdown due to Covid-19 gave enough time to write his contemplations of years in volume-1 of this series. He wrote this book with pure intentions of sharing the knowledge and invoking the basic idea of 'understandings' with unprejudiced minds and open hearts, in order to help the society.

Akhil has been giving motivational speeches at various corporate events and on various other occasions. Akhil's belief in working on "Leave Positive Footprints" is the driving force to part in various philanthropic and environmental activities.

info@akhilrajendra.com
www.akhilrajendra.com

CONTENTS

1 - MY NOTE AND MY GRATITUDE

1.1- My Note

Decoding 'The God' and 'The Religion' is written in a format that is easy to read and understand, whether the reader is a 13-15 years old or 50-60 years old, a 10th grader or a post graduate. It caters to everyone in a simplified manner with the idea of not asking one to 'believe', but to understand, feel, introspect, explore and know the answers. Each line is written matter-of-factly, which will make readers feel like this is something they already knew, they already had inside them and will hopefully encourage you to look within.

Evolution plays a huge part here in this era, so to help you enjoy the experience of your individual journey into the divinity, I do not need to make you wear the coloured spectacles of belief and dogmas. In today's time, there already is so much information, and awareness, so many studies on physical and medical education, which are progressing every passing day. Social and judicial systems also get upgraded in most democratic countries, so I do not need to write a lot about it; that would be like rewriting an existing software as religion in a book or this series of books. All I need to do here is remove the coloured lens of belief and dogma, and open all eyes to understanding, giving accessibility to upgrades and updates of existing software, with the right intent, taking the human brain as the hardware.

I have refrained from taking the name of any religion, sect or belief, or of their respective preachers, so that opportunists do not mislead people from the right context and intent of this book for the sake of their greed, lust, personal gains and agendas.

"God is too big to be fixed in one religion or to be confined in its one book."

#AkhilRajendra
#LeavePositiveFootPrints

1.2- My Gratitude

With great humility I convey my gratitude to all my gurus, those whom I remember and those whom I might miss out but contributed to this book; to all those who taught me how to read and write in my childhood and to everyone who contributed to my academics, trainings and skills. To my parents, family, and friends, and to the profound scriptures and their respective writers, who gave me knowledge and enlightenment. To the people, animals and things that gave me conditioning and experiences along my journey of life.

I would like to mention three names here with my special thanks - Swami Vivekananda, Charles Darwin and Stephen Hawking. These great personalities have been my inspirations during my journey through the world of knowledge and spirituality.

Hope you truly enjoy reading the Volume -1

"God is too big to be fixed in one religion or to be confined in its one book."

#AkhilRajendra
#LeavePositiveFootPrints

2 – WHO ARE WE 'THE HUMAN KIND'?

2.1 - The Universe

Our universe is an ancient and vast infinity, and is expanding farther and faster every day. This expanding universe is filled with stars, galaxies, planets and more, plus a veritable buffet of invisible stuff like dark matter and dark energy. The cosmological description of the development of the universe starts with the big bang. According to this time line, space and time (known to us all four dimensions of X, Y, Z and time) emerged together through a black hole about 13.799±0.021 billion years ago. The energy and matter initially present became less dense as the universe expanded. After the initial accelerated expansion called the inflationary epoch - and the separation of the four fundamental forces known as gravitation, electromagnetism, and the weak and strong interactions - the universe gradually cooled and continued to expand, allowing the first subatomic particles and atoms to form. Dark matter gradually gathered, forming a foam-like structure of filaments and voids under the influence of gravity. Giant clouds of hydrogen and helium were gradually drawn to the places where dark matter was most dense, forming the first galaxies, stars, and everything else seen today. It is possible to see objects that are now more than 13.799 billion light-years away because space itself has expanded, and it is still expanding and evolving today. This means that objects which are now up to 46.5 billion light-years away can still be seen in their distant past, because in the past, when their light was emitted, they were much closer to Earth.

2.2 - The Solar System

Fast forward approximately one-third of the way along the timeline of the universe, 4.5672±0.0006 billion years ago. One of the ongoing cosmic events in the universe was the accretion from the solar nebula, containing gas, ice particles and dust, including primordial nuclides from a giant interstellar molecular cloud. At this time, a gravitational collapse occurred, as the gravity pulled swirling gases and dust in to itself, which then began to spin and flatten into a disk. Then the planets grew out of that disk, and that was how our solar system formed. Most of the collapsing mass collected in the centre to form the Sun, while the rest of the mass flattened into a proto-planetary disk out of which the planets, moons, asteroids and other bodies formed.

The four smaller inner planets, Mercury, Venus, Earth and Mars, are terrestrial planets, primarily composed of rock and metal. The four outer planets are giant planets, being substantially more massive than the terrestrials. The two largest planets, Jupiter and Saturn, are gas giants composed mainly of hydrogen and helium. The two outermost planets, Uranus and Neptune, are ice giants composed mostly of substances with relatively high melting points compared with hydrogen and helium (called volatiles) such as water, ammonia and methane. Meanwhile Pluto's orbit is moderately inclined relative to the ecliptic (over 17°) and moderately eccentric (elliptical). The solar system settled into its current layout about 4.5 billion years ago.

2.3 - *The Earth*

Planet Earth formed with the rest of the solar system, 4.54 Billion years ago, the third planet from the Sun. Like its fellow terrestrial planets, Earth has a central core, a rocky mantle and a solid crust. Earth's gravity interacts with other objects in space, including the Moon, which is the Earth's only natural satellite. The Earth orbits around the Sun in 365.256 solar days, a period known as a sidereal year. During this time, the Earth also rotates about its axis 366.256 times. The Earth's axis of rotation is tilted with respect to its orbital plane, resulting in the seasons on Earth we experience. The gravitational interaction between the Earth and the Moon causes tides that help in stabilizing the Earth's orientation on its axis. Earth is the densest planet in the Solar System, and the largest of the four rocky planets.

The Earth's atmosphere and oceans were formed by volcanic activity and outgassing, which was supported by late heavy bombardment . The water vapour from these sources condensed into the oceans, augmented by water and ice from asteroids, proto planets, and comets. Atmospheric greenhouse gases kept the oceans from freezing when the newly forming Sun had only 70% of its current luminosity. By about 3.5 billion years ago, the Earth's magnetic field was established, which helped prevent the atmosphere from being stripped away by the solar wind, unlike the other three rocky planets. This became an ideal environment for life to grow in, with respect to the temperature, atmosphere, water and the seasons.

2.4 - Origin of Life on Earth

Before I go into the details, take a moment. Just think about life a few hundred years ago, when electricity or engines weren't invented yet. The inventions and evolution that have taken place in the last 300 years have made such a drastic difference to human life, and here we are talking about changes that happened over a period of four billion years.

Life on Earth depends largely on carbon and water, which Planet Earth had in plenty to nurture it. Carbon provides a stable framework for complex chemicals to form, and can be easily extracted from the environment. A half billion years after the formation of the Earth, chemical reactions led to the first self-replicating molecules about four billion years ago. Independent emergence of life on Earth, known as a biogenesis or an organism's ability to produce offspring that are very similar to itself, began. Their ability to feed and repair on their own evolved, and external cell membranes which allow food to enter and waste products to leave, but exclude unwanted substances, formed. Photosynthesis allowed life forms to harvest the Sun's energy, the oxygen produced accumulated in the atmosphere and due to the interaction with ultraviolet solar radiation, formed a protective ozone layer in the upper atmosphere. The evolution and formation of smaller cells of microorganisms within larger ones resulted in the development of complex cells called eukaryotes. True multi-cellular organisms formed, as cells within colonies became increasingly specialised. Aided by the absorption of harmful ultraviolet radiation by the ozone layer, life colonized on the Earth's surface. This was the epoch when the evolution of sexual reproduction in eukaryotes started as meiosis and fertilization. There is ample genetic recombination in this kind of reproduction, in which descendants receive their genes from each parent, in contrast with asexual reproduction where there is no recombination. Bacteria also exchange DNA through bacterial conjugation. However, conjugation is not a means of reproduction and is not limited to members of the same species, as bacteria also transfer DNA to plants and animals.

2.5- Various Species on Planet Earth

Approximately one trillion species currently live on the Earth. Out of these, only about 1.8 million have been named and even fewer than about 1.6 million are documented. These currently living species represent less than one percent of all species that have ever lived on Earth. This means hundreds of trillions species have lived been eradicated from the Earth by now.

About 750 to 580 million years ago, multi-cellular life forms significantly increased in complexity, and there was a mass extinction of various species. When the continental land masses broke up, the movement caused a volcanic spree that also released massive amounts of carbon dioxide into the atmosphere, which resulted in a short period of global warming. Eventually the lava began to cool and harden. Over a period of tens of millions of years, the weathered rocks separated sufficient carbon dioxide to plunge the Earth's climate into an extreme ice age.

This was Snowball Earth – a period that began around 715 million years ago and held the Earth in its icy grip for over 120 million years. This crushing catastrophe saw the rise of one of the most incredible steps in evolution: the evolution of the first animals, and a dramatic flourishing of life.

Around 540 million years ago, a host of exotic creatures appeared, including giant woodlouse-like creatures known as trilobites. Dominant life on Earth went from single-celled bacteria to exotic multi-cellular creatures.

2.6- *The Human Evolution*

About 66 million years ago, an asteroid impact triggered the extinction of the dinosaurs and other large reptiles, but spared the smaller mammals. Creatures that resembled shrews and rodents remained, and mammalian life diversified over time. Just over four to three million years ago, an African ape-like animal in the class of Australopithecus, identified as Orrorin, the early species of Homininase, evolved the ability to stand upright. This facilitated the use of tools and encouraged communication. This provided the nutrition and stimulation needed for a larger brain with more neurons, and led to the evolution of Homo Erectus. Early Homo Erectus fossils found in East Africa have been traced back to around 2 million years ago. About 200000 years later, the species had migrated to modern-day Georgia, at the border between Eastern Europe and Western Asia. As Homo Erectus is a direct ancestor to Homo Sapiens, human beings are one type of the several living species of great apes. Humans evolved alongside orangutans, chimpanzees, bonobos, and gorillas. All of these primates share a common ancestor that can be traced back about 7 million years ago. Homo Sapiens, the first modern humans, evolved from their early hominid predecessors between 200,000 and 300,000 years ago. They developed a capacity for language about 50,000 years ago. The first modern humans started to travel across and beyond Africa about 70,000-100,000 years ago, and continue to travel across the world today. The development of agriculture, controlling and using other animals and the eventual dawn of civilization, led to humans having a major influence on the Earth and nature that continues to this day. The human body, brain, senses, emotions and control of all neurotransmitters are conditioned for the known to us four dimensions - X, Y, Z and time - like any of the other species. The variety is the one of beautiful basic characteristics of the entirety. Every creature of this entirety may look like or be categorised as the same, but they are never exactly the same. If we sow one seed, its tree or plant will not have exactly same size, colour or even qualities. All their fruits and seeds will also have a difference. The same applies to the human race, its nature, characteristics, behaviours and minds.

3 – THE GOD

3.1 - The idea of 'God'

The idea of 'God' in the human mind came during the process of evolution, when the brain of one particular specie (Homo Sapiens) started developing more rapidly on this planet, compared to the brains of other species.

The human brain subsequently developed into a mind capable of choice and decision making, and it began asking questions about its existence. 'Who are we?' 'Where were we before'? 'What am I doing here?' 'Where we shall go after?' 'Who controls all this and why?' These questions and more gave birth to the idea of an entity understood as God.

I truly do not mean here that 'God' was not there before this evolution or will not be there after its eradication, but not the one installed through religious beliefs, which are like operating systems or software in minds, considering brain as its hardware.

3.2 - When It's Said 'God' Is One

When it is said that "God is One", it sounds like a good thought, with the intent of resolving, if not solving, all unanswered questions. The evolved human mind sees God as a distinct unit and other things like the sun, moon, self, animals, water, air and everything else as other separate units. The evolving mind then works out the other details of God - form/ formlessness, his all supreme nature - supported by imagination, the perks and rewards of heaven, punishment of hell, and many more 'hypotheses'. This is where things go wrong, when the same idea of God being one, which was intended to unite people, becomes the biggest divisive factor and oversight the 'God' himself. Continuously evolving minds need better answers. They deserve logical answers and patience, not a reply merely within the confines of belief. Here starts a journey towards the answers by understanding, knowing, and experiencing God.

3.3- What is "God"?

Most people have a corner in their evolving minds that believes in someone or something far superior than their evolved capabilities. They fantasise about having a mental or emotional connection with it, which is also what gave birth to the idea of superheroes. During ancient times, these superheroes were kings, angels, deities, saints, God's son, God's messengers, and mythological heroes. These superheroes used to be the saviours, during any powerful, emotionally charged or desperate situation of emergency, they would save the lives of everyone and do wonders against all negative, evil forces and injustice.

However, where are those superheroes today? When all of human civilization is struggling hard against this historic Covid-19 outbreak, the global pandemic. (As on the day of writing this chapter, 5th April 2020).

To this day the stories and their characters are the same, but they live now in comics, TV shows, video games and films with contemporary screenplays, music, visuals and detailing as Superman, Spiderman, Batman, Wonder Woman, Captain America and many more.

Here is my important revelation, when people claim to love and adore God, then why have an oppressive approach towards God and confine him in a religion or in a book. God is in his entirety, me (my name, given by my late loving father, also means so), you, and everything around you is God, infinity is God and so is what lies beyond it, matter is God and energy is God, and God is the source of this same matter and energy. Four fundamental dimensions as X, Y, Z and time created during the big bang, there are many more dimensions, as in physic the primary dimensions are said as mass, length, time, temperature, electric current, amount of light, and amount of matter, these known and unknown to us dimensions are also God, and God was there before the big bang, God was the big bang, and God will be many more similar and different big bangs and such or different cosmic events.

Nothingness in nothing is God and God is everywhere, the pure and with and without form one. He/ she (just to complete the sentence) is us, and also our creator, the past, the present and the future; he/she is to be loved, worshiped, he/she is the love and worship and God is the only the lover and worshiper, omnipresent and omnipotent, the divinity and the entirety. Every single atom with their electrons, protons, and neutrons, the vacuums, the electromagnetic forces within them, are God and what God is made up of.

Every person, animal and plant, living and dead is God. Each specie to be born and live and that has already lived and died is also God, and the same applies with non-living things, the destroyed ones were God, the existing ones are God and the ones to be created naturally or by hand, existing and waiting to be destroyed are also God. The same continuation is also God.

Sound is God, light is God, their speed, travelling and destination is God and are created by God. As God is without a beginning, middle or ending, God is the complete cosmic manifestation and its creator, and the creator of the moving and the immobile.

This planet, its moon, its solar system, its sun, this galaxy having billions of solar systems, stars and many more cosmic things in it, is God. The whole universe with billions of galaxies and a lot more cosmic things in it, is God and beyond the universe is also God. He/ she is the creator of God itself and himself/ herself is also God, God is this book and also God the one who is reading it.

Ancient scriptures about 17,000 to 20,000 year old, recognize God as almighty and merciful, the source of all strength, bearing the burden of the whole universe.

Thus the sages (spiritual masters who spent years thinking and contemplating continuously evolving minds, having basic questions about our existence and of this whole cosmic manifestation) said we shall worship him only through pure love, as this is what we ultimately want in return from God. The universe functions on frequencies, and responds to similar frequencies at the wavelengths they are emitted at. Ancient sages reached to the idea that he is to be worshiped as one beloved, dearer than everything, in this life and the next with many names and forms by understanding the universal unity of the divinity. When we love and worship him, we should also love and worship his creations, be it fellow human beings or animals; that is why animals and plants were considered up in devotion, due to their holy (useful to human kind) properties. Above is the reason it is said that whatever is given to universe comes back in form of Karma.

It was believed that there is a God, who is Brahmana (not the one now understood as the caste, but in Sanskrit *brah* means larger and *atman* means the soul, literally meaning larger soul), the ultimate truth. He has two aspects of his nature. The *Nirgun Brahman* is without any attributes and *Saguna Brahman* is the creator, the preserver and the destroyer of the universe. The thirty-three types of celestial being are his perceptions. This led to a misconception among the progeny that lives today, that there are thirty-three crore (0.33 billion) gods. In Sanskrit, *Koti* means type/ classification as well as *infinity* (as in *UchchKoti* – high category and *NimnaKoti* - low category). In reality, the idea in the scriptures was that there are 8 Vasus, 11 *Rudraas*, 12 Suns, 1 *Indra* and 1 *Prajapati*, totaling 33 demi gods. Just like humans are more evolved than other species, it was considered that demi gods are superior to humans, considered to be appointed by the Supreme God to oversee the functions of creation. In fact, it was a delegation of authority and division of labour in the celestial world. The demi gods are considered to not die natural deaths, instead they get dissolved into the super consciousness at the time of dissolution of the universe. Demi gods are worshipped because of their proximity to God and because of their functional authority. The entire process of creation, preservation and destruction are part of God's cosmic play. The concept of *Trimurti* symbolises these aspects of the God Maya i.e. illusion and we are all small characters in this play, assigned a definite role to play. External symbols are used in worship with respect to their particular symbolic properties, so that the physical image always reflect in the mind during prayer. Though it goes unsaid that the material image/ idols don't confine God within them, for centuries they have kept intact the idea of holiness, purity, truth, love, benevolence, and kindness. Hence, they helped in keeping minds fixed on whom to pray to attain divinity. Images, idols, temples and books are only supports, help if you will, for spiritual infancy, but on and on it must progress. Scriptures have said that the truth and God can be realized in many ways, and every individual is empowered to worship in any way one desires. The path may be different but the goal is common. Vedic scriptures did not deride, demean, or disparage other methods of prayer, on the contrary they recognise their purity of intent and respect the same. The conclusion from the scriptures was that God can be felt and worshiped at all times. There is no particular day or time fixed for his worship, as God is not part-time fascination, but a full-time obsession. This Vedic *mantra* gives one of finest explanations of

God: -

Om Puurnnam-Adah Puurnnam-Idam Puurnnaat-Purnnam-Udacyate

Puurnnashya Puurnnam-Aadaaya Puurnnam-Eva-Avashissyate ||

Om Shaantih Shaantih Shaantih ||

Meaning: *Aum!* That is infinite, and this (universe) is infinite. The infinite proceeds from the infinite. (then) taking the infinitude of the infinite (universe), It remains as the infinite alone. *Aum!* Peace! Peace! Peace!

In knowing, understanding, praying, meditating or worshiping God, one should keep in mind that God is not a different unit or body than him/ her and we create our vibrations in the universe to be answered with similar frequencies. We now know that our universe works on frequencies of all matter and energy. This will further be elaborated in chapter 6.2.

Some oceanic species, reptiles, mammals and birds also have been found to have some notions about the idea of God, similar to the mankind went through during its evolution of its mind of having the questions about its existence and this whole creation.

4 – THE RELIGION

4.1 - The Operating System

Spirituality in human life continued during the process of evolution, as the human mind came up with more questions, more ideas, more curiosity and everything else the human mind is capable of, along with the challenges to their basic instincts of survival and reproduction. When modern day human at various locations on earth had newly evolved, In order to satisfy their curiosity, humankind formed the idea of a resolution instead of a solution. That was the birth of the idea of having religion, the software having the coding of the road towards spirituality and no human being would socially function without again for a very long time.

During our childhood, playing outside in the evening till after dark, the Sun would set and we would be drenched in sweat, but we wouldn't give up on a very interesting game. Our mothers would get tired of calling us back home, but we would still want to play for some more time. Frustrated with this daily calling back and pulling back exercise (different human nature applies here, as few children were needed to be pulled back), mothers would often say and establish in our minds that there is a monster living behind that far wall of the ground. He has sharp fangs, red eyes, long nails and always in search of children's blood after sunset. Often mothers would present some fabricated evidence that the monster knows how to fly in the clouds or it broke that rock or cut off those branches of that tree, or the same monster has already taken away a few children of neighboring localities. So we believe our mothers without any skepticism.

We would get so frightened, we would sleep early hugging our mothers and won't have courage to go to the washroom in the night without lights. The next day we would avoid walking alone even from the far of that wall in the daytime also and wind our games at sunset before it gets dark. We were also given perks after completing a given task or finishing our homework or after scoring good marks in exams - chocolate boxes, toys or fulfillment of a dream wish of latest sports cycles.

Looking back at the faded images of our childhood memory lanes, we see that the park or ground's far wall had no monster behind it. That broken rock and cut tree was the work of the municipality people; now those chocolate sweets are just extra calories and that dream wish of the latest sports cycle does not draw our attention. We just smile in understanding that It was all just like scarecrow effigies on a farm with the pure love and intentions of our beloved mothers in our childhood, as they were concerned for our safety betterment and well-being. We are evolved now from children to grown-ups, our priorities have changed, so have our dreams and our ambitions; also our choices for happiness, satisfactions of our ambitions and dreams have changed.

Similarly, the detailing of the heaven we have now may not be ultimate thing of our journey, as we are evolving continuously along with the universe. With humility, I compare here the pure intentions of those saints and preachers, who in all the religions gave the idea of 'Heaven', its glory/perks, and 'Hell' and its pains/punishments.

These saints and preachers had to make efforts by making rules, giving more detailing of these destinations as heaven, hell and judgement day etc. by leaving no skepticism with a very noble intention of making people follow the path of righteous and honesty and lead a social life for their betterments, well-being and keeping them together with compassion, as per the conditioning, lesser literacy percentages and understanding levels of the people of those eras of different human nature.

As per the illiteracy rate data of 2016, 14% world's adult population, which is more than a billion, still does not know how to read and write their own names even. In 1820, it was 88% of adult population who did not know how to read and write their own names, 500 years ago it was even lesser and a 1000 years ago it was even more lesser. This percentage of illiteracy also varies in different regions, countries and continents at different times. How one can expect them to be educated enough to understand with right meanings, the knowledge or books written after great contemplation and studies at different locations, conditions and times? Unlike today, there was no education system of different books for different grades with considerations of student's understanding level.

So there has been one common idea behind to help the progeny remember and follow the given preaching and guidelines and not to deviate from the path of righteousness, compassion, humanity and spirituality. The religious rules and practices were given with the preaching and detailing of heaven, hell and judgement day. This is very similar to the way in scriptures the characters had divine and supernatural properties with very interesting stories, having all the emotions and contents to hold interest with even entertaining quotes in detail. Many temples and shrines (older than 100s and even 1000s years) were made on the basis of these engaging stories from the mythological scriptures, having the symbolic postures of their characters, so that people should not forget the path of righteousness, compassion, humanity and spirituality and keep remembering and practicing the same with right spirit and honesty for long for their betterment and well-being.

And above both have been helpful also for so many centuries, but as I said before that our minds are evolving and need more understanding than just being in locked thoughts with the idea of 'belief', which is causing conflicts among faiths. A confined in 'belief' human mind, where thinking job is done by someone else, is actually against the basic principle of the evolution. This is the reason we see that the human minds which are free from the confines of 'belief' are not only bigger achievers but are also biologically more evolved now, in every field of human life in every part of the world, let that be engineering, medical science, astronomy, art, business, discoveries, literature and just any and everything.

Here I come up with my revelation. It is just a figment of the imagination that God lives in a heavenly destination, which is a wonderland-like platform somewhere in the sky, having a gate and lot more fictitious things inside it, issuing its passes through a faith or its book's preaching, or the practice or following of the path of any religion. All one needs, a sense of understanding, and one can feel, experience and actually live in heavenly depth without dying or killing others, as explained before that life before birth was God, life itself is God and life after death is also God. There was no such time when God was not there and there will not be any time when God will not be there, as time is just one dimension and there are many more dimensions.

4.2 - *Dharma*

In Hindi, religion is called *Dharma,* a Sanskrit word that the English language actually does not have a literal word for. *Dharma* is the search for the truth of our existence, our purpose and our goal. It is much more than religion, which is ambitious and very limited in its approach. Religion is sectarian, whereas *Dharma* is universal. The purpose and objective of religion is to keep its respective followers together with rigid rules and dogmas, through preachings by an individual in certain circumstances, situations, and eras for the people of that region as per their conditioning and with references of available scriptures, knowledge and information, supported by cogitation.

The Sanskrit word *'Sanatan'* in English means 'eternal', *Sanatana Dharma* is a way of life designed through evolution to ensure the continuity of humanity on this earth and provide the entire population with spiritual sustenance.

Dharma endows you with guidance to attain the ultimate truth.

Dharma was revealed through meditation, study, and experience by great sages in the past. Their revelations are recorded in the scriptures, mainly in *Vedas.* Those sages did not mention their names in the scriptures, so as not to give a beginning or ending to it like the true nature of God the supreme being. *Vedas* are considered as the accumulated treasures of spiritual laws discovered by different people during different times. Just as gravity existed before its discovery, so too were the laws that govern the spiritual world. The rishis propounded the *Vedas* and some of the very great *rishis* were women. The *Vedas* taught that creation is without beginning and ending.

The Vedic system was intended to help and prod the individual to attain spiritual truth by personal experiences in their journey of divinity. It did not lay down dogmatic rules to follow, but insisted that the individual shall draw his own methods. It allowed full freedom to find answers to spiritual questions without restrictions. The Vedic system knew and explained that there are many ways to find the truth. *Ekam sat vipra bahuda vadanti* means 'There is only one truth (or true being) and learned people call it by many names.'

Vedic knowledge enlightens the same truth in many ways. There is no bar on seekers and they can study and probe other religions too. The personal relation of an individual with God is absolute. Vedic study explains that all living beings are spirits living in mortal bodies, and the bodies shall die but the soul shall not, profoundly said in the *Gita,* "Him the sword cannot pierce, him the fire cannot burn, him the water cannot wet, him the air cannot dry."

The *Vedas* teach that the purpose of the soul is to attain enlightenment and reach *Moksha,* to merge with the *Brahman,* (the larger soul, the absolute truth). The soul is always pure, but as the destructible body covers the soul, it gets contaminated. The demands of the body are mundane and material. When we allow the body to satisfy all its demands, we in a way weaken our souls. It is said that the nourishment for a powerful soul is positive and good thoughts, which lead to good *Karma,* thus we lose our direction and goal. This deviation from the set path brings misery and misfortune; it changes the destined course of our life.

Dharma highlights that there are six negative characteristics prevent man from attaining moksha, which are 1- Lust or desire *Kama,* 2- Anger *Krodha,* 3- Greed *Lobha,* 4-Fascination or attachment *Moha* –delusory emotional attachment or temptation, 5-Arrogance *Mada* – ego or arrogance and 6-Jealousy *Matsarya* – envy, jealousy. These bind the soul to the cycle of birth and death and keep it confined in this material world (confines of Maya or relative existence). Especially the first three.

Dharma tells us how to keep our soul pure and profound, and helps us in achieving our goal. Our past actions determine our present and the present guides the future. There must have been causes and past actions before our births to make us miserable or happy. The human soul is eternal and immortal, perfect and infinite. The soul will go on evolving, reverting from birth to birth and death to death. *Dharma* advises that we refuse to call ourselves sinners, the *Vedus* further explain that the soul is divine, only held in the bondage of matter. Perfection will be reached when this bond breaks and we attain mukti - freedom from our bonds of imperfection, and freedom from death and misery.

This bondage can only be broken through the mercy of God, and this mercy falls on the pure. Hence purity is a condition of His mercy, he reveals himself to the pure heart. The pure and the sinless see God even in this life; then and only then is all the crookedness of the heart corrected. Then all doubts cease, and they no more remain ordinary mortals. They become divine and knowledgeable. This is the very centre, the very vital concept of *Dharma*.

Thus the whole objective of the system of human life is by constant struggle to become perfect, to become divine, to reach and see God, and this reaching God and seeing God means becoming perfect even as the God is perfect, which constitutes our religion. When he attains perfection, he lives a life of bliss, having obtained the only thing in which man ought to have pleasure, namely God, and enjoys this bliss with him.

We believe that after death the soul becomes free but its further journey depends on the performance in the previous life. Good performance is rewarded and the bad deeds are punished. It is said that in reincarnation, we consider human life precious as the soul has reached this stage after many births and rebirths though many species even on this planet and others. In the human stage the goal is nearer to the truth. But if we indulge in bad *karma*, we are pushed back and have to begin our journey afresh.

Dharma accepts all the religions from the lowest fetishism to the highest absolutism. The many attempts of the human soul to grasp and realise the infinite are each determined by the conditions of its birth and association, and each of these marks are a stage of progress. Every soul is a young eagle soaring higher and higher, gathering more and more strength, till it reaches the glorious Sun. Unity in diversity is the plan of nature, and *Dharma* recognises it.

Nowhere in *Dharma's* scriptures is it said that only we shall be saved and others will not. *Dharma* has no place for persecution or intolerance. In Dharma, It is practised as polytheism, is more likely to be aligned with monism than is monotheism. The reason is those who worship the Divine in many names and forms can more easily recognize Universal Unity than those whose One God rejects all other deity forms. It recognises divinity in every man and woman, and it instructs to aid humanity to realise its own nature, the divine nature of the soul. *Dharma* prods 'faith', which has its beauty, peace, patience, respecting and valuing others, and experiencing one's own journey of divinity.

4.3 – Bigotry and Its Root

As said before that in 'belief', someone else has already done the thinking by leaving no space of any skepticism. In the sectarian religions, mostly created in order to follow a path keeping their flock together, it is installed as software. This may or may not have helped them keep their flock together, but surely identifies others as 'them', and here on the bigotry starts bourgeoning.

One religion says, "We are right in our 'belief'."

The rest of the world says, "Good to know."

The same religion says, "We are the 'only' ones who are correct in our 'belief' and the rest of you are 'wrong'."

Rest of the world says, "Okay, stay happy as long as it pleases you."

The same religion says, "If you don't believe in our 'belief', you are non-believers of God. Making you accept our 'belief' by doing just anything is one of the agendas in our mission and is the gate-pass to heaven, having written the name of our belief's preacher on its gate, in our language only."

Some fanatics from the same religion blindly believe that killing the people of other faith or to those who are resisting the acceptance of their 'belief' is a sacred duty that they owe to God and your fellow followers and ensures a gate pass to heaven, which doesn't exist anywhere in the sky as a wonderland kind of platform, having a gate and many fictitious things inside, as I explained in details above in 4.1.

These religious radicals believe that their religion is the

ultimate truth and they must establish their religion as superior wherever they can by giving their interpretation of their religious texts, by overlooking that the religious text was written in certain situations of that era, as per the conditioning, lesser literacy percentages and understanding levels of the people of different natures.

The people who do not believe in their radical notions are left with no choice but 'resistance' for their freedom, including spiritual freedom. This resistance is in or out of proportion, at times democratic or autocratic. Clever and desperate opportunists take advantage of the situation by leading the same to the 'chain-reactions' of communal disharmony and religious fights, for the sake of their personal gains and agendas.

Such opportunists have always been there and will always be there in every region and era in some or other form. For example, the business of one of the most powerful and influential lobby of the world, international arms and ammunition manufacturers and their brokers, depends upon war and war like situations. Other example is certain media houses; they have their marketing teams, sales targets, always in desperate situations of repaying to their bankers or delivering results to their investors from promoting their policies and agendas, their targeted and selective approach of making a news, at times fabricating a news story to a level of purposely portraying the victim as accused to ignites a flare. They make it more engaging with their skilled editorial by adding more fuel, supported by drama, punchlines and background music for the sake of their personal agendas and professional objectives.

But again, as an example here, if our computers and phones have a software malfunction and do not match the newest need, we do not abuse the phones; we do not kick, break or throw it out of the window. Rather we must upgrade the software when an upgrade is available, the same computers and phones become so much easier to use with their software upgrade. This is one example I give, whenever people talk of dealing such above mentioned fanatic extremists with similar brutality. My stance here is that if we really want to fix the problem, then instead of being part of the problem, we need to work to solve the problem by finding its orientation. I believe and opine here that there is surely a need to make these people aware of the damage they are doing, and give them a more holistic picture of the world beyond their irrational blindness and now ignoring this need by others is not a right thing. This sounds like a daring job, in other words their dogmatic religious beliefs and practices have their terror in people's minds. I would prefer the same be done through an open dialogue with understanding, with open hearts, unprejudiced minds and by not letting the opportunists convert the same in to the reactions for their individual, political, personals gains and agendas.

Such religion based fanaticism has been there in every part of the world for centuries. It is truly painful to see so much of the loss of human lives and blood, including the killing of innocent unarmed men, women and children for the chase of the gate-passes of heaven. This is perhaps one of the reasons for me to write this book and the following series, which ensures not just the gate passes, but all of heaven, within the lifetime itself, without dying or killing others and it surely does. And believe me, sooner or later people have to agree with this fact that God is too big to be fixed in one religion or to be confined in its one book, sooner would most certainly be wise and better for human mind.

4.4 - Death and After Death

Death is a very simple process everyone has to undergo, whether it is human beings, animals or plants. Death is the end of life in a living creature when all biological activities and living activities stop, including of the mind and the senses. The usual signal of death in humans and many other animals is that the heart stops beating and cannot be restarted. This can be caused by many things. All living things have a limited lifespan, and all living things eventually die. Most people have seen the dream of their own death at least once. Every time, the moment they die in their dream, people end up waking up. Why? Because our conscious, subconscious or even unconscious mind does not have any information about post-death experiences, which proves that stored thoughts in the brain are associated to this body, from the senses of evolved human mind.

The *Vedas* explained that the death is end of the life in the body but not the end of the whole journey of the soul, as every soul is a circle of which the circumference is nowhere, but whose centre is located in the body. Death means only the change of this centre from body to body. Nor is the soul bound by conditions of matter. In its very essence, it is free, unbound, holy, pure and perfect. But some way or the other, it finds itself tied down to matter, and thinks of itself as matter. Every living body is the representation of the soul or 'Atma', which is part of God. The soul in indestructible and the body gets destroyed, not eliminated but rather gets dissolved in the five elements of matter - earth, water, fire, air and space, from which the body was formed. When the body gets merged in all the components it's made up of, then only *karma* in the form of energy is transmitted.

4.5 - Karma

Karma is complex and difficult to define in few paragraphs. A *Karma* theory considers not only the action, but also the actor's intentions, attitude, and desires before and during the action and even the thoughts too are self-generated and create energies as *Karma*. As explained in chapter 3.3 that the universe functions on frequencies, Universe responds to similar frequencies at the similar wavelengths they are emitted at, is the reason it is said that whatever is given to universe comes back in form of *Karma*.

A common theme of the theories of *Karma* is its principle of causality. One of the earliest associations of *Karma* to causality occurs in the *Brihadaranyaka Upanishad*. For example, at 4.4.5–6, it states:

Now as a man is like this or like that, according to as he acts and according to as he behaves, so will he be. This means a man of good acts will become good, a man of bad acts, bad; he becomes pure by pure deeds and bad by bad deeds.

Here they say that a person consists of desires, and as is his desire, so is his will; and as is his will, so is his deed and whatever deed he does, that he will reap.

Good *Karma* has a good effect on the doer, while bad *Karma* has a bad effect. This effect may be material, moral or emotional — that is, one's *Karma* affects one's happiness and unhappiness.

The effect of *karma* need not be immediate; it can be later in one's current life, and in some schools of thought, it extends to future lives.

The effects of one's *karma* can be described in two forms: *phalas* and *samskaras*. A *phala* (literally, fruit or result) is the visible or invisible effect that is typically immediate or within the current life. In contrast, *samskaras* are invisible effects, produced inside the doer because of the *Karma*, transforming the agent and affecting his or her ability to be happy or unhappy in this life and future ones. The theory of *Karma* is often presented in the context of samskaras.

Karma seeds habits (vāsanā), and habits create the nature of man. *Karma* also seeds self-perception and perception influences how one experiences life events. Both habits and self-perception affect the course of one's life. Breaking bad habits is not easy; it requires conscious karmic effort. Thus psyche and habit link *Karma* to causality in ancient Indian literature. The idea of *Karma* may be compared to the notion of a person's character, as both are an assessment of the person and determined by that person's habitual thinking and actions.

Every action has a consequence, which will come to fruition in either this or a future life; thus, morally good acts will have positive consequences, whereas bad acts will produce negative results. An individual's present situation can thereby be explained in reference to actions in his present or in previous lifetimes. *Karma* is not itself reward and punishment, but the law that produces consequences. Good *Karma* is *Dharma* and leads to punya (merit), while bad *Karma* is considered *Adharma* and leads to pāp (demerit, sin).

This is so because the ancient scholars of India linked intent and actual action to the merit, reward, demerit and punishment. A theory without ethical premise would be a pure causal relation; the merit or reward or demerit or punishment would be same regardless of the doer's intent. Ethically, one's intentions, attitudes and desires matter in the evaluation of one's action. Where the outcome is unintended, the moral responsibility for it is less on the doer, even though causal responsibility may be the same regardless. The concept of *Karma* thus encourages each person to seek and live a moral life, as well as avoid an immoral life.

The concept of reincarnation or the cycle of rebirths (samsāra), has been intensely debated in ancient literature of India; different schools of Indian religions consider the relevance of rebirth as either essential or secondary, or unnecessary fiction. *Karma* is a basic concept, rebirth is a derivative concept. *Karma* is a fact and so the transmission of Karmic energies is sure. In the Bhagwat Gita, the *Karma* philosophy is explained brilliantly, addressing all questions of all emotional states of the mind in ideal situations in order to explain the details of God.

Rebirth, or samsāra, is the concept that all life forms go through a cycle of reincarnation that is a series of births and rebirths. The rebirths and consequent lives may be in different realms, conditions or forms. The *Karma* theories suggest that the realm, condition and form depend on the quality and quantity of *Karma*, every living being's soul transmigrates (recycles) after death, carrying the seeds of Karmic impulses from the life just completed, into another life and lifetime of *Karmas*. This cycle continues indefinitely, except for those who consciously break this cycle by reaching *Moksha*. Those who break the cycle reach the realm of gods, those who don't continue in the cycle.

Considering this, we need to introspect calmly and see clearly the immense compassion, non-violence, kindness, humility and understanding not only among humans but also for other species needs to be practiced as all have one common ancestor on earth, when the life on it started evolving a four billion years ago, during the Eoarchean Era.

5 – THE SCRIPTURES

Scriptures, available since the last 10,000 to 15,000 years are considered as sacred revered texts or Holy Writ, of the world's faith and spirituality. Scriptures comprise a large part of the literature of the world. They vary greatly in form, volume, age, and degree of sacredness, but their common attribute is that the devout regard their words as sacred.

Scriptures differ from literary texts because they are a compilation or discussion of beliefs, mythologies, ritual practices, commandments, laws, ethical conduct, spiritual aspirations, and for creating or fostering a religious community. The relative authority of religious texts develops over time and is derived from the ratification, enforcement, and its use across generations.

The terms 'sacred text' and 'religious text' are not necessarily interchangeable. Some theistic religious texts are believed to be sacred because of the 'belief' attached to them, that the text is divinely or supernaturally revealed where in non-theistic religions they are considered to be the central tenets of their eternal *Dharma*. Many religious texts, in contrast, are simply narratives or discussions pertaining to the general themes, interpretations, practices, or important figures of the specific religion, the modes of worship and other sectarian scriptures.

5.1 - Vedas

The *Vedas* are said to be the most ancient, profound and complete scriptures known and available to human civilisation. *Vedas* provide more information about spiritual domain, the characteristics of God and our relationship with the supreme. The *Vedas* are a large body of religious texts originating in ancient India. Composed in Vedic Sanskrit, the texts constitute the oldest layer of Sanskrit literature and the oldest scriptures. The Vedic hymns themselves assert that they were skilfully created by *Rishis* (sages), with immense thought process and by investing a lot of knowledge, hard work, experiences, experiments, research, pondering and meditation.

Veda means knowledge, to know. It is said that the *Vedas* contain profound secrets of the world, their origins, and descriptions of the spiritual world, which a man can enter by the power of his will and consciousness.

The origin of *Vedas* is unknown, and while researchers date it back to 3500 years, new evidence dates it back 10,000 years. In any case, these scriptures are ancient. There are some references about their origin in the *Upanishads* and *Puranas*. Vedic literature is of two types - *Shruti* and *Smriti*. *Shruti* means the one that is heard. The *Shruti* scriptures were revealed to ancient *Rishis* (sages) through meditation. These invaluable *Shrutis* were handed down to us by oral traditions. It was secret knowledge taught only to a selected few, written down much later, sometime before 1500 B.C.

Each *Veda* is divided in four parts, namely the *Mantra, Brahmna, Aranyaka* and *Upanishad*. While the *Mantra* contains mantras, the *Brahmana* part contains information about rites and rituals.

The *Aranyakas*, meaning written in the forests, deals with the philosophical background of the various rituals. The fourth part of the *Vedas* is called the *Upanishads*. The *Upanishads* are the books of deep spiritual knowledge and gives meaning to *Vedas*. The *Smriti* scriptures are an explanation of Shruti, written by mortal humans to make the general populace understand the scripture.

The *Shruti* includes four *Vedas* and according to tradition, Vyasa (full name Vyasa Krishna Dwaipayana) is the compiler of the *Vedas*, who arranged the four kinds of mantras into four *Samhitas*(collections). There are four *Vedas*: -

1.   Rigveda, a book of mantras that consists of 1017 hymns. Many *Rishis* composed them, and each mantra has immense power and profound meaning.

2.   Yajurveda, a book of rituals. It exemplifies the procedure for cleansing of mind and attaining eternal bliss, and helps us to dissolve into the supreme being.

3.   Samaveda, is actually a smaller version of the Rigveda, with the important mantras of Rigveda recorded in song versions. It gives meaning to the Rigveda.

4.   Atharvaveda is again a set of hymns, free from the strict regimen of rituals. It also talks about the life of people during Vedic times.

5.2 Upanishads

Each *Veda* has been sub-classified into four major text types – the (mantras and benedictions), the *Aranyakas* (text on rituals, ceremonies, sacrifices and symbolic-sacrifices), the *Brahmanas* (commentaries on rituals, ceremonies and symbolic sacrifices), and the *Upanishads* (texts discussing meditation, philosophy and spiritual knowledge). Some scholars add a fifth category – the *Upasanas* (worship). Each *Vedas* has a set of *Upanishads*, which explains the *Vedas*. Upanishad literally means to sit near. In Vedic times, the knowledge of the *Upanishads* was given to students of highest merit, whose sincerity, intelligence and devotion had been verified by their guru. Since the knowledge was imparted when students sat down near their teachers and listened to them the word Upanishad came in vogue.

In the serene solitude of the forest, the learned Gurus inscribed the sacred philosophies of the *Vedas* in their disciples. The knowledge of the *Upanishads* is considered *Brahmvidya*, the ultimate science of self-realisation. It deals and explains with the help of simple stories, social and spiritual aspects of our life including subjects such as the nature of true knowledge, the ideal human conduct, the practice of *Yoga*, the concept of *Karma* and the incarnation of the soul.

The prominent figure in the *Upanishads* is the sage Yagnavalkya. He taught the great doctrine of neti-neti-neti, an important concept that states that truth can found only through non-recognition of the self.

Although there are 108 *Upanishads*, only eleven are considered to be the most revered: *Isha, Kena, Aitareya, Katha, Prasna, Mundaka, Mandukya, Taittriya, Chandogya, Brahd-aranyaka* and

Svetasvatara. Of these 108 *Upanishads,* 10 pertain to Rigveda, 51 to Yajurveda, 16 to Saamveda and 31 to Atharveda.

Six sentences sumrise the sacred and pious teachings of *Upanishads,* known as Mahavkyas (Which literally means Great-Sentences). They are: -

1-   *Aham Brahmasmi:* I am part of the supreme soul; our own self is the reality and he resides in me.

2-   *Ayam Atma Brahma:* The self is the supreme soul/God, I am the divine and there is no difference between me and him

3-   *Tat Tvam Asi:* That thou art, we are the ultimate, we are what we think

4-   *Prajnanam Brahma:* Our inner most intelligence is the supreme of all and has the capacity to merge with the absolute

5-   *Sarvam Khalvidam Brahma:* The universe is the supreme being God, one including the self; everything is within me and I am within everything

6-   *So Ham:* I am that, I am that which resides in every breath

5.3 - Other Important Scriptures

One truly needs to understand the difference between Scriptures and Mythology. Smriti scriptures contain the works of various individuals, who have interpreted the *Vedas* according to their intellect. Smriti means memory. It is a sacred intellectual literature, meant for the purpose of human welfare. Smriti scriptures include five distinct groups of writings. The scripture has its logic and guidance; over the millennia of its development with evolution, *Dharma* has adopted several iconic symbols that are imbued with spiritual meaning, based on either the scriptures or cultural traditions.

Mythological history is made up of highly entertaining stories, situations, catharsis, emotions and characters. They have been helpful in continuing the interest of the progeny in the scriptures and in passing on customs and values through books, temples, statues, customs and rituals. Mythology made it easy to convey and carry forward the knowledge of *dharma* in the progeny we see today, regardless of the education levels of the people, their age groups, region or language even with the help of the value added in the projection of its characters with divinity.

Symbols from *Dharma*, as per Vedic science, are not mere signs but have important significance. Vedic science advises accordance of high importance and reverence to such symbols.

1-   *Itihasa* means history and contain mainly the Ramayana and Mahabharata.

2-   *Puranas* contains the events and stories of the incarnations of Vishnu.

In today's era, it is called mythology. Nevertheless, it is a divine narration with divine characters, and includes information and knowledge in the form of very interesting stories that make them entertaining and memorable. They pass on the same knowledge and information of cosmology, rules and values for life, rituals, instructions on spiritual knowledge, with strengthening our character and lead us to truthful and the divine ways of life.

Though the exact number of *Puranas* are not known, 18

Mahapuranas and 18 *Upapuranas* are accounted for. The *Mahapuranas* are *Brahmpurana, Padampurana, Garudpurana, Vishnupurana, Shivapurana, Bhagwatpurana, Naradpurana, Markandeyapuarana, Agnipurana, Bhavishyapurana, Varahpurana, Kurmapurana, Matsyapurana* and *Brahmadpurana.*

3- *Dharma Shashtras* are law books dealing with social and human life and includes *Manu Smriti* and *Yajnavlkya* Smriti among others.

4- *Agmas* deal with topics like the codes of temple building, image making and the modes of worship and sectarian scriptures.

5- *Drashanas* are manuals of philosophy and the main schools are as follows:

    a- *Nyasa* and *Vaisheshika*

    b- The *Sankhya* and the yoga

    c- The *Mimansa* and the *Vedanta*

6- Ayurveda, the science of health

7- *Dhanurveda*, the science of archery, a part of the *Vedas*. There were a total 4520 titles of Vedic literature, but today we have only 256 titles.

8- *Shiksha* - The science of proper articulation and pronunciation

    a- *Chandas* - The study of the Vedic metres

    b- *Vyakaran* - The rules of grammar

c- *Nirukta* - The explanation of difficult and complex Vedic words

9- *Jyotisha* - The study of astronomy and astrology

10- *Kalpa* - The process of symbolic sacrifice (not the killing of animals or anything) and associated rituals

Lord Vishnu the protector has 10 avatars (incarnations)

1- Matsya (The Fish)

2- Kurma (The Tortoise)

3- Varaha (The Boar)

4- Narasimha (The Lion Man)

5- Vamana (The Dwarf)

6- Parashurama (The Lumberjack)

7- Lord Rama

8- Lord Krishna

9- Lord Buddha

10- Kalki

As per the scriptures, Kalki, the last and final incarnation of Vishnu is still expected. If I am the one then it is a letdown for those expecting super hero qualities and fun stories from this incarnation, as here it's all about rational knowledge by invoking the 'Understanding'.

5.4 - *The Bhagwad Gita*

The Bhagavad Gita means 'the song of God'. The Bhagavad Gita is the best known and most famous of Indian texts, with its influence. The Gita's call for selfless actions without being instrumental in making our actions bear fruit, also let not our attachment be instrumental to inaction, has been inspiring many great personalities across the world for a spiritually healthy life. Bhagwad Gita tells to keep our soul happy by feeling and knowing the divinity inside with inner happiness, which is the great service to God and shows a process of keeping purity in thoughts, which shall help in purifications of our *Karma*.

While reading Bhagwad Gita, one must focus on understanding the right context and meanings, as it is said in Gita itself during a dialogue between Sanjay and Dhritrashtra, that it's not always but during this conversation with Arjuna, Ultimate God's divine revelations are coming from the mouth of Krishna, where Krishna is like a medium in a particular situation.

Bhagwad Gita's 700 verses are structured into several ancient Indian poetic metres, with the principal being the shloka (*Anushtubh Chanda*). Each shloka is a couplet, thus the entire text consists of 1,400 lines. Each shloka has two-quarter verses with exactly eight syllables. Each of these quarters is further arranged into two metrical feet of four syllables each. The metered verse does not rhyme. While the shloka is the principal meter in the Gita, it does deploy other elements of Sanskrit prosody. At dramatic moments, it uses the *tristubh* meter found in the *Vedas*, where each line of the couplet has two-quarter verses with exactly eleven syllables. The Bhagavad Gita is comprised of 18 chapters.

Chapter 1 (46 verses) is about wrong thinking is the only problem in life

Chapter 2 (72 verses) is about Right knowledge is the ultimate solution to all our problems

Chapter 3 (43 verses) is about Selflessness is the only way to progress and prosperity

Chapter 4 (42 verses) is about how every act can be an act of prayer

Chapter 5 (29 verses) is about renouncing the ego of individuality and rejoicing in the bliss of infinity

Chapter 6 (47 verses) is about connecting with the higher consciousness daily

Chapter 7 (30 verses) is about living what you learn

Chapter 8 (28 verses) is about never giving up on yourself

Chapter 9 (34 verses) is about Valuing your blessings

Chapter 10 (42 verses) is about seeing divinity all around

Chapter 11 (55 verses) is about having enough surrender to see the truth as it is

Chapter 12 (20 verses) is about absorbing your mind into His highness

Chapter 13 (35 verses) is about detaching from *maya* and attach to the divine

Chapter 14 (27 verses) is about living a lifestyle that matches your vision

Chapter 15 (20 verses) is about giving priority to Divinity

Chapter 16 (24 verses) is about how being good is a reward in itself

Chapter 17 (28 verses) is about how choosing the right over the pleasant is a sign of power

Chapter 18 (78 verses) is about letting go and moving to union with God

The theology of the Gita elaborates the nature of God, the nature of the self, and the nature of the world, *Brahman-atman*, the way to God, *Karma yoga, Bhakti yoga, Jnana(Gyana) yoga,* synthesis of the *yogas, Raja yoga,* asceticism, renunciation and ritualism, *Dharma,* allegory of war, war and duty, and *Moksha.*

The Bhagwad Gita is a part of the epic Mahabharata (chapters 23–40 of *Bhishma Parva*). It is set in a narrative framework of a dialogue between Pandava prince Arjuna and his guide and charioteer Krishna. At the start of the *Dharma Yudhha* (righteous war) between the Pandavas and Kauravas, Arjuna is filled with a moral dilemma and despair about the violence and death the war will cause in the battle against his own kin. He wonders if he should renounce and seeks Krishna's counsel, whose answers and discourse constitute the Bhagavad Gita. Krishna counsels Arjuna to "fulfill his Kshatriya (warrior) duty to uphold the *Dharma*" through selfless action. The Krishna–Arjuna dialogue covers a broad range of spiritual topics, touching upon ethical dilemmas and philosophical issues that go far beyond the war Arjuna faces. Some also see Krishna as the first motivational speaker in human history.

Numerous commentaries have been written on the Bhagavad Gita with widely differing views on the essentials. *Vedanta* commentators read varying relations between the Self and the Brahman in the text; *Advaita Vedanta* sees the non-dualism of Atman (soul) and Brahman(universal soul) as its essence, whereas *Bhedabheda* and *Vishishtadvaita* see *Atman* and *Brahman* as both different and non-different. The setting of the Gita in a battlefield is interpreted as an allegory for the ethical and moral struggles of human life.

The Bhagavad Gita presents a synthesis about *dharma*, theistic devotion and the yogic ideals of *moksha*. The text covers *jnana*, *bhakti*, *karma*, and *Raja Yoga* (spoken of in the 6th chapter) incorporating ideas from the *Samkhya* Yoga philosophy.

6 – THE MEDITATION

"In prayer, you speak to God and in meditation, you converse and listen to God"

6.1- The Meditation

The English word Meditation means to think, contemplate, devise, and ponder, which is understood as the meaning of Sanskrit words *'Yoga'* and *'Dhyana'*, They are practised to realize the union of one's eternal self or soul, one's *atman*, through Vedic Science. Yoga is too vast and challenging to define. It is a practice where an individual uses a technique such as mindfulness, or focusing the mind on a particular object, thought, or activity, to train their attention and awareness. This helps them achieve a mentally clear and emotionally calm stable state, as practices vary both between traditions and within them. Ancient Vedic meditative techniques have spread to other cultures where they have also found application in non-spiritual contexts, such as business and health. In today's era, meditation is therapeutically in use with the aim of reducing stress, anxiety, depression and pain, and increasing peace, self-confidence and well-being. Meditation is under research internationally to define its possible health (psychological, neurological, and cardiovascular) and other effects and is prescribed by medical practitioners for the treatment of anxiety, depression and many other mental illnesses.

Meditation has been practiced since antiquity in numerous traditions, often as part of the path to enlightenment and self-realization. The earliest records of meditation, come from the

Indian traditions of Vedantism.

The oldest documented evidence of the practice of meditation is wall art in the Indian subcontinent from approximately 5,000 to 3,500 BCE, showing people seated in meditative postures with half-closed eyes. Written evidence of any form of meditation was first seen in the *Vedas* around 1500 BCE. In India, the tradition of Guru and *Shishya* has been around for ages, where students were sent to *Gurukuls* mostly in the forests to live and learn under a learned experienced teacher. During this time and for centuries before, all learning and knowledge was passed on by word of mouth. Almost all the scriptures talk of meditation in some form or the other. Hence, we can safely assume that meditation was also an integral part of the knowledge that the Gurus were teaching their students, and all this was done orally, which is also what makes it very difficult to tell how old meditation really is.

In *Advaita Vedanta*, this is equated with the omnipresent and dual Brahman (bigger soul). In this dualistic yoga school and *Samkhya* (one of the six *astika* schools in Vedic system of education mainly related to the yoga and Dhyanam) the self is called Purusha, a pure consciousness separate from matter. Depending on the tradition, the liberating event is called *moksha, vimukti* (detachment).

The earliest clear references to meditation in Vedic literature are in the middle *Upanishads* and the Mahabharata (including the Bhagavad Gita). The earlier *Brihadaranyaka Upanishad* describes meditation as "when having become calm and concentrated, one perceives the self (*atman*) within oneself".

One of the most influential texts of the classical Vedic yoga is Patañjali's yoga, a text associated with yoga and *Samkhya*, which outlines eight limbs leading to **kaivalya** (aloneness). These are ethical disciplines (*yamas*), rules about physical postures (*āsanas*), breath control (*pranayama*), withdrawal from the senses (*pratyāhāra*), focus of mind (*dhāranā*), meditation (*dhyana*), and finally death (*samādhi*).

Later developments in Vedic meditation include the compilation of Hatha Yoga (forceful yoga) compendiums like the *Hatha Yoga Pradipika,* the development of *Bhakti yoga* as a major form of meditation, and *Tantra.* Another important Vedic yoga text is the *Yoga Yajnavalkya,* which makes use of *Hatha Yoga* and *Vedanta Philosophy.*

I shall elaborate the intent and context of this above paragraph in the next volume/volumes of this series.

6.2 - Prayers

Energy is not always in the form of heat, but in frequency. The universe responds to your frequency. It doesn't recognize your personal desires, wants or needs; it only understands the frequency at which you are vibrating. For example, if you are vibrating with the frequency of fear, guilt or shame, you are going to attract things of similar vibrations to support your frequency. If you are vibrating in the frequency of joy, love and abundance, you are going to attract things to support that frequency. It is like you have to be tuned into the radio station you want to listen to, just like you have to be tuned into the energy you want to manifest in your life.

Here is the revelation regarding praying to the deities, the goddesses, regardless of real or the scripture's characters, super heros, saints or a particular place or object. "I am the reason for a prayer, and I am the one who is praying. I am the prayer, I am the one in form and without form to whom the prayer is offered. I am the one, the wish, asked or offered in the prayer. I am the place, the structure from where prayer is offered; I am the result of that prayer. If the prayer is answered, then I am the one who will be rejoicing but if the prayer is rejected by me, then again I am the one who will be disappointed. The joys and disappointments are also me, like all other virtues, feelings, emotions and just everything of this universe and beyond, which is also me."

When the heart is full of faith, devotion and love, it sings. The words of such mystical songs are prayers. Prayers are a submission of requests, made to person/s or God, in whom the seeker has immense faith, trust and love. It may be a call of distress, need, or simply a song of love, expressing or eulogizing the loved person. Prayers are our desires or dreams that we aspire to fulfill by expressing it to our beloved God in whose ability and mercy we have immense faith. The prayers are directed to someone with whom we can communicate without any fears or reservations. Simply putting, any communication with God is a prayer. It may be in the form of prose or poetry too.

Now it has been scientifically proven that, physiologically and spiritually, prayer has positive effects on us. It is a powerful tool to purify us, provided it is offered with devotion, faith, love and a positive approach. But before we pray it is necessary to calm the mind through meditation.

As per the scriptures, broadly speaking prayers are of two types - one seeking a favour, and the second out of love and pure devotion. Our ultimate objective shall be the second. The daily chanting of prayers invokes the God within ourselves and brings us closer to him in total surrender. We are able to see divinity within ourselves, and all around us. It is said that the continuous chanting of prayers creates an energy field that prods and helps the seeker to achieve his desired goals. Moreover, during the prayer, we converse with God. We submerge ourselves in his thoughts, we are able to concentrate and meditate upon him. We vent our feelings without any inhibitions, which gives us instant relief from distress or misery. During prayer, God becomes our friend and a companion. It is said in *Vedas* that while praying one should concentrate at the point between your eyebrows (*Ajna Chakra*), *Vedas* all the three *nadis* i.e. *Ida, Pingala,* and *Sushumna* converge at this point. These nadis are positive, negative, and neutral energy flowing in our body, which can be aroused by practicing *Kundalini* yoga. By concentrating on the *Ajna Chakra*, both the sides of the brain get activated and positive energies emanate. One of the finest prayers, chanted during most occasions is:

Tvameva Mata cha Pita Tvameva, Tvameva Bandhu cha Sakha Tvameva,
Tvameva Vidya Dravinam Tvameva, Tvameva Sarvam Mama Deva Deva.

Tvameva Mata: You are my Mother, the supreme one without a second, who nourishes me with Divine Love and graces my life with self-respect (the perception of myself as the soul-self, and not just as a body, mind, intellect, or ego).

Cha Pita Tvameva: And you are my Father, the supreme being who protects me by raising my consciousness and transforming my mind into a receptacle of the divine; who instils in me the sterling qualities of divine consciousness.

Tvameva Bandu cha Sakha Tvameva: You are my true relative

with who I am eternally related in *Atman* (Soul-self) with *Paramatman* (the Supreme Self), and you are my best friend, my eternal companion and dearest confidante who will never leave me.

Tvameva Vidya: You are the divine wisdom, the essence of everything I know, everything I am learning, and everything I do not know but seek to understand and realize (actualize).

Dravinam Tvameva: You are the highest wealth (*Lakshmi*) and the bestower of all the best things in this life and the next. You are the source of everything good and the bestower of all resources we require for our physical sustenance and spiritual enlightenment. You are the wealth of wisdom and the gift of liberation (*Moksha*).

Tvameva Sarvam Mama Deva Deva: You are all-in-all; you are everything to me. You are the core of my being, the heart of my heart, the source of my self, the soul of my soul, the ultimate reality devoid of duality and partiality; indivisible, immutable, immaculate, the ultimate knower and the absolute perfect incomparable supreme being.

6.3 Mantra

In Sanskrit, mantra means something that protects the mind.

MAN= MIND, TRAI= TO PROTECT

It is believed that the very mention of the word mantra invokes sacred, divine and pious feelings in us. We are awed by its sacrosanct and heavenly connotations. The inherent art and master of it, is present in all of us.

Mantras are considered as a spiritual tool to awaken the dormant power present within us. It is a tool to attain ultimate bliss. A mantra reveals to us the true purpose and objective of life. The intricacies and mysteries of life become easy to comprehend with the aid of a mantra. By knowing a mantra, everything becomes known. Nothing baffles us. The few secrets and sacred words unveil a new world. The world we crave to belong to.

According to the *Vedas*, the creation of the world began with a being i.e. Naraayana. From Naarayana as a being, came the mind; from the mind came desire, and from desires, the will emanated. The vibrations of the will created a word. The word was 'AUM', also known as *Naada Brahma*. The word *aum* created a sound that created an energy field, and everything else followed the *Naada*, the sound. That sound is a frequency, the cause for all creation.

It is said that Sound can be created by vocals or by striking different things, but the first ever sound was created was unstruck, the sound of *Anant Naada* (the sound of infinity) which still pervades through the whole universe. The *spanda* (vibration) of that sound holds all the *bhutas* (elements). Our essential nature is consciousness and bliss, but due to our obsessive preoccupation with the mundane matters, we allow and develop many unwanted elements or feelings to disturb our peace. Then the restlessness creeps in and the consciousness and bliss is shattered. Once again, we crave and yearn for peace and try to reconnect ourselves to the *Ananda Naada*, which always vibrates in our consciousness. Mantras are considered as a vehicle for such a ride. Mantras are very ancient and are more than 10,000 years old. Mantra were conceived by the sages in their deep meditations and studies, and the sages have transmitted these mantras to the progeny for crossing the illusion of this external world and attain the pure bliss.

It is a divine speech to dispel the darkness clouding our consciousness. Mantras are considered the most important tools for clearing and cleansing the minds, mantras is said to help break up our thought and desire patterns, which keep us enslaved to insatiable materialistic urges.

All day long, something or the other goes on in our minds. It is a mindless cacophony. It encompasses our thought- field and serves to divert our attention. It is usually not possible for us to silence such a mind. It is believed that It is however always within our power to chant a mantra and instil some sense and discipline into the mind. If we do this regularly, the mantra gradually replaces the background noise of our mind and ceases to resist our puerile, churlish and goalless intentions. This is said to be the use and purpose of the mantra, they help us change the nature of our mind, as we are bound to have some or other emotional problems in life. The solution is to practice the mantra rather than engage in trying to solve fruitless and tiresome riddles of such emotional waves. The egoistic mind causes mental suffering, the spiritual mind through mantras is the cure has been the idea.

6.4 -Mantra Meditation

Using mantras in meditation was a given technique of repeating a pure sound vibration to achieve inner peace, which delivers the mind from its inclinations and illusions. Chanting (*Jaapa*) is the process of repeating a mantra. This act of chanting mantras is mantra mediation. The purposeful continued chanting creates a field, which is believed to help us attain eternal bliss.

The true purpose of a mantra is to boost and empower the mind, hence when we vibrate our mind with the right sounds (*dhwani*) and correct words with meaning our mind becomes calm, peaceful, purposeful and concentrated. It gets the ability to read and analyse thoughts with the right perspective. When this transformation of speech occurs, we begin to find the profound meaning in few simple words. Even the sound of some simple syllables such as *aum* soothes us and opens the vistas of new horizons and new meaning. They break down the barrier of language and take us to the universal state of communication, which is silence and peace. It affects our *karma* of the present and the past. It can also help with material concerns, necessities and emotions.

Besides being a sound pattern with meaning, a mantra is also considered an energy composed of certain frequencies that have a pattern of their own. These frequencies and the positive effects generated by them influence the nerves spread around our internal organs. The positive response generates and influences the two sides of our brain.

Just as yoga or any ancient pious practice requires a special *yoga shakti* to facilitate it, the practice of mantra does too. For a mantra to work it must have at least the purity of purpose and total concentration of the mind. A medium or help is required. This could be the grace of a teacher that aids the growth, or the power of an initiation or our own insight, and a connection with the guide who dwells within us (*Upaguru*).

The mystical transcendental potency for spiritual realizations is already present within everyone. All we have to do is lift the cover. It must, however, be uncovered by genuine spiritual process. In the preliminary stage of chanting, the practitioner experiences a cleansing of consciousness, peace of mind, and relief from unwanted drives and habits. As one starts chanting, he realizes and sees the original, spiritual existence of the self.

6.5- Mantra Chanting

As per scriptures, *Vedas*When the mantra chanting (*Japa*) is done with discipline and purpose, positive results are a foregone conclusion. There are many ways and methods explained for chanting. But one should adopt a method which is soothing and easy.

Fast chanting exhausts the mind, heart and breath, and relaxation comes only after the chanting is over. Slow chanting relaxes the mind, heart and breathing while the chanting is going on, but it is good only when the chanting is done individually.

Apart from the variations while chanting the mantras, they can also be chanted in the following four ways: -

1- *Vaikhari-* The chanting begins loudly and gradually the pitch lowers.

2- *Madhyama-* One chants silently within oneself. Even the lips do not move.

3- *Pashyanti-* Chanting in a trance.

4- *Para-* Chanting that one enjoys, and where the mind finds divine peace.

Chanting slowly induces a state of trace, the sound creates space which creates sympathised and positive sounds, and these sounds influence the entire nervous system. More space (*Akasha*) is produced when sound echoes. The echoing quality of sound increases with the increase of *Akasha*.

The constant hammering of sound kneads the flesh, nerves and cells of the body. It affects the petals in the *chakras* of the human body. Mantras are of two types; long chants and short seed syllables. The longer mantras are chants, like the Vedic verses. The shorter root mantras have a deep universal meaning. We can use them according to their energy even if we do not understand the language. The longer chants depend more on our interpretation of their meaning and on our intentions.

The serious aspirant can combine the practices of chanting, with *Namsasmarana*, prayers and meditation, *bhajan* singing and seeking out the company of good people (*Satsang*), in order to develop *Sadhana*, and attain purity with perfection.

6.6- Importance of Language

God needs no language, words or sound as even thoughts contain frequencies. God is the language, word, sound and frequency himself.

Sanskrit is one of the most ancient languages and considered as one of the utmost perfection, in human civilisation as not just a human language but the reflection of the vibratory power of cosmic intelligence into the human mind, which allows us to link to the universal Consciousness and repository of eternal knowledge.

It is the root of a great number of languages, including the main languages of Europe, India, many of the languages in the Persian region and Middle East. It is the language of the

Vedas and of the great sages. It is the only language where the sounds are made in the mouth. Sanskrit has more words for the divine and more precise terms for defining consciousness and meditative experience than any other language. It is considered the language of the higher mind and thereby gives us access to the law and vibratory structure, and to the powers of these domains.

According to Vedic studies, Sanskrit alphabets produce a spiritual effect. It is an energy-based language and is immortal, pure and potent. Hence, it is believed that a mantra said in Sanskrit is more effective. Each one of the Sanskrit words contains important and significant inner meaning, and both the meaning and sound have potency.

6.7- Requisite to Say Mantras

Vedic science only promotes peaceful mantras, done without any desire for reward, power, or attachment. These mantras free one from diseases, psychological problems, fearful illusions and worldly and environmental troubles.

As per the scriptures, there are ten types of mantras for various objectives, but I am refraining from mentioning their types even here, as at times people get carried away with temporary emotions and reactions and use the mantras to satisfy their temporary desires. However, their karmas are carried forward for their redemption. So coming back to the peace mantras, though there are no definite set of rules or methods of mantra chanting, a broad set of do's and don'ts have been handed over to us by the serious practitioners. All said and done, one can develop a method with which one is most comfortable.

The following are the time tested set of rules for mantra chanting, guided by the scriptures.

1- One can set a pledge of 7, 9, 11, 15, 21, or 40 days or more. One is free to either obtain a mantra from his/ her guru or decide and select a mantra, considering a guru in any of the deities as per the deity's characteristics.

2- During the pledge period, the mantra should be recited. It is advisable to recite a set of the same mantras every day. Arrive at the ideal mantra by trial and error method.

3- One can begin the chanting (*Japa*) after taking a bath, or if for some reason one cannot take the bath, then he/she can wash their hands and feet or sprinkle water on their head and say the following Mantra before the beginning of chanting:

Om apavitrah pavitro sarvaavashtam gatopi va;
ye samret pundarikaassham sa bbahyabhyantarah shuchich | |

(I may be in any state, but by reciting the pious name of Lord's incarnation Vishnu, I become clean.)

4- Mantras can be chanted in a mala, having 108 seeds in it, at least one *mala* in one sitting. For beneficial results, it is said to chant 16 malas in one sitting, twice a day. One should chant at a particular time, preferably at *Brahm Muhurt*, 2 hours before sunrise and two hours before it sets.

5- One should complete the mantras without any hindrance and interruption in a single place which is pleasant and clean. This should preferably face North or East, but there is no hard and fast rule on this. Sit in a comfortable position but do not lean on supports unless physically incapable.

6-   One should use *malas* of *tulsi* or *rudraksha* containing 108-plus-1 beads for counting. Move the beads with your middle finger and thumb; *meru* is the main knot of the mala. One never crosses it, but turns the Mala around and continues the next round of chanting with the same.

7-   Think of your *Aradhya* deity or some saint or your favourite picture of god. During the pledge period one should consume *Satvik* food and maintain physical, mental and spiritual purity. One should preferably chant the pledge from the same place without any interruption, but if the course is long, and then one can recite while walking, driving or through a journey. After any interruption, one should recite the *mahamantra, Om Namah Shivaya* five times and resume the same.

8-   One should be vocal in the beginning to master the verse, and the pronunciation, but later on transcend to a deep hum. One also needs to work on the correct pronunciation of words.

9-   The ending of mantras is *Namaha* if you are 28 years or younger and have not had your Saturn return, and *Swaha* for those older than 28 and have had the return. The energy rises to the solar plexus *chakra* after 28 or so, and is signified by *Swaha*. There are a few mantras (*Durga*) that also use *Namaha*.

10-   One should try to stay celibate during the pledge period. There is also a mantra asking the divinity for help in staying celibate, while concentrating on a bowl of water/

milk 21 times and then consuming the same after chanting. The mantra is:

Om Namo Bhagwate Mahaablale Parakramaaya
Manobhilaasheetam Manah Stambh Kuru Kuru Swaha: | |

11- One should refrain from revealing his/ her pledge to others. Even in the case that it is revealed, there is no need to panic. Your inner energy can be more powerful than anyone else's on the same pledge. It all depends on your frequency. Before they pledge a mantra, one should stay calm for a while and chant *aum*, and after the completion of the chanting period, one can extend it if desired so for benefits. After daily chanting, one should not leave their seat immediately, rather they should do it gradually with calmness.

12- And one should remember to complete the mantras with
"Om Shanti, Shanti, Shanti"

6.8 What is Beeja Mantra?

It was sages with their study, research, mediations and through inner revelations through hundreds of years have put into Vedic scriptures the *Beeja* mantras. These are basically words without any particular meaning to them, but have tremendous sounds and vibrations. Most of these are highly secret, mysterious, and potent mantras, consisting of one or more *Beeja* syllables. These can be chanted individually, or in conjunction with other *mantras.*

Shrim (pronounced *Shreem*) is the mantra of beauty, grace, prosperity, and abundance and relates to the Goddess. It brings about the fulfilment of all wishes.

Eim (pronounced as *Aym*) is the sound of wisdom and relates to Saraswati, the Goddess of knowledge. It increases our powers of attention, concentration, reason and contemplation. It is also the sound of Guru the spiritual teacher, and brings into play the learning power of the higher mind.

Klim (*Kleem*) is combined with the other mantras for effect and potency.

Dum, (*Doom*) is a mantra for protection, and *Krim* (*Kreem*) for *Shakti*. *Gum* (*Sum*), and *Glaum* are used for removing obstacles.

Chakra Beeja Mantra
(for *Kundalini* yoga followers)
Om Lam Vam Yam Ram Yam

Ram is the sound of divine light, grace and protection. It gives strength, peace and compassion.

Hum (pronounced as *Whom*) is the sound of divine wrath. It destroys all negativity. It is a sound of Shiva sound and a sound of *Agni,* the divine fire, through whom all the other Gods or divine powers can be invoked.

Om (*Aum*) emerged from the mouth of Brahma, the creator and the creation, before he created everything else. *Om* is the essence of all the mantras, the highest of all mantras, the divine word or Shabda Brahmna (the creator and creation). *Om* is also said to be the essence of the *Vedas*. Even one ignorant of *Vedas* can chant it. *Om* is the sound of the infinite. It gives power to all the mantras. Hence, most of the Mantras begin and end with *Om*. *Om* itself clears the mind for meditation. It is the word of the past, present and the future. It is the beginning and the end of everything. Everything dwells within it, it is the rhythm of life. *Om* in itself is a profound and powerful mantra.

It consists of three sounds; the vowel 'a', the vowel 'u' and the nasal 'm' sound. Hence, it is sometimes written as *Aum* (Or Alm as in Arabic, L is often pronounced as U when it comes before another consonant).

There are three sound patterns of om, relate to the respective states of waking, dream and deep sleep and to the three Gunas of Rajas, Sattas and Tamas. Brahma, Vishnu and Mahesh are in a threefold role as creator, sustainer and destroyer of this universe. In the *Vedas*, Om is the sound of the Sun, the sound of light. It is the sound of an ascent, an upward movement and uplifts the soul.

Om Kaaram Bindu Samyuktam Nityam Dhyaayanti Yogniuh
Kaamandam Mokshadam Chiava Omkaaraaya Namo Namaha

The sages meditate upon the sound of om and its *bindu* (dot). We offer obeisance to the sound of *om*, which has the power to fulfill our desires and grant *moksha* to us.

6.9 - Soham

Soham is the natural sound of breathing. '*So*' is the sound of inhalation and 'Ham' is sound of exhalation. If we breathe deeply and listen to the sound of our breath, we hear these sounds. In Sanskrit the root '*Sa*' from which *So* derives itself, means to hold, to have power, to be. It gives inspiration and power. *Ham* means to leave, abandon, and cast out, hence to expel or exhale. If we eliminate the consonants from *soham*, we get '*Om*'. *So*, is Shakti and *Ham* is Shiva; through them we balance the God and the Goddess within ourselves. *Soham pranayama* is natural deep breathing with attention to the sound of these mantras.

Hakaren Bahiryati Sakaaren Vishet Punah
Hansa Hansa Ityayam Mantra Jivo Japati Sarvada

The body always chants *Hansa, Hansa. Hansa*, which is the combination of *Ham + Sa*, the process of exhaling and inhaling air. If we concentrate on our breathing, the mantra benefits will accrue. In Sanskrit, Soham also means I am that cosmic being. He resides in me; I am part of him too.

6.10 - Aham Brahmasmi

This means I am the part of the Brahma (Greatest Soul, the creator and the creation), the supreme being. The eternal power dwells within me.

6.11 - Worship

Pooja is the ritual associated with the chanting of various mantras, and the recitation of *shlokas* and prayers with offerings to God and the deities. It helps us to concentrate and focus our attention on the deity. The physical act and handling of rituals gives us the satisfaction and feeling that we are actually serving a divinity. God never demands anything from us. Mortal human beings try to please God in different ways, of which one of them is offering pooja with pure love, sincerity and devotion. The feeling behind it is symbolic and conveys the feelings of a devotee to his deity, which goes to God the entirety, as revealed in the chapter 6.2.

In *Manas pooja*, the devotee does not need to offer any tangible or materialistic thing. He or she is able to concentrate fully on God as there are no rituals associated with this offering. It is economical and considered effective too. The devotee is free to conceptualise and dream to any extent to please the divinity, with no rituals and no materials, only sheer joy. The following is prayed during this *pooja*:

Oh, dear God! I am offering to you the earth in the form of sandal wood paste, sky as the flower, air as the scented smoke, fire as the light, nectar as the sweet and all the best treatments available in nature.

Sanvatsara pooja, which is New Year *pooja*, is celebrated as per the Vedic calendar on *Chaitra Shukla Pratipada*, with cosmological significance. Since 1st January is observed as New Year's Day, many perform the *pooja* on that day too, which starts with a pledge in Sanskrit:

|| *Mama Sakutumbasya Saprivarasya Swajan Parijan Sahitay Va Aayuraarogyaisharvaryaadi Sakala Shubh Phalottaraabhivriddhyrth Brahmana Disanvatsara Devataaman Poojunmaham Karishye* ||

This is followed by various recitations of *mantras*, *shlokas*, and offerings, pleasing all the deities, appeasing the planets and *nakshatras*, apologising for sins, and then distributing the *prasada* with the appropriate rituals.

7 – THE SYMBOLISM

The scriptures retain their logic and guidance over the millennia of their evolution. *Dharma* has adopted several iconic symbols with spiritual meaning based on either the scriptures or cultural traditions. The exact significance accorded to any of the icons varies with region, period and denomination of the followers. Over time, some of the symbols, for instance the Swastika, have received wider association while others like *Om* are recognized as unique representations of *Dharma*. Other aspects of *Dharma* are covered under the terms murti for icons and *mudra* for positions of the hands and body.

Symbols from *Dharma*, as per Vedic science, are not mere signs, but have important significance. It advises to accord high importance to such symbols and revere them.

7.1 *ShivaJyotirLinga*

It is very important to understand the logic behind symbols. Let us understand what *ShivaJyotirLinga* means. In Sanskrit, Shiva means the entirety in a dot, *Jyoti* means light and Linga means form. Thus, *ShivaJyotirLinga* stands for *Bindu Prakash Swaroop,* light in the form of a dot. The *ShivaLinga* we see in temples is symbolic of the *ShivaJyotirLinga* made up of rock as flame of lamp in its stand and to draw our attention with prominence and respect from within.

7.2 – Om

Om is considered as the first word, the sacred word, representing a connection with God who is within.

7.3 - Swastika

It is the second most important symbol, the Swastika. It is not a word or a letter, but a line-drawn picture. The Swastika symbolises the eternal nature of God, for it points in all directions, thus representing his omnipresence. The term Swastika is made of two Sanskrit words - Su meaning good and *Asati* meaning to exist/ happen. When combined, it means 'May good prevail'.

7.4 – Poornakumbha

An earthen pot full of water, with fresh mango leaves and a coconut on top of it, is placed before the chief deity or by the side of the deity. The pot symbolises Mother Earth, water and leaves symbolise all life sustained, and the coconut represents consciousness.

7.5 - Naiveda

Naiveda is the food is offered to God as a symbol of our ignorance, and after the offering, we accept it back as *Prasad*, accepting that God has cast his benevolent spell on our ignorance and purified it. When we share it with others, in essence we are sharing this newly acquired knowledge.

7.6 - Saffron

The colour saffron pervades all facets of our life. It is the colour of *agni* and represents the purity of God. The saffron flag symbolizes *agni* and valour. The people who have renounced material life wear a saffron robe to announce their renunciation.

7.7 - Lotus

The lotus flower grows in swamps, but remains unaffected by the filth. The flower symbolizes non-attachment with the filth of the material world and shines above *Kaama* (lust), *Krodha* (wrath), *Moha* (attachment), *Lobha* (greed) and *Ahankar* (ego/pride).

The creator of this universe Brahma is also depicted as being seated on a lotus, symbolising that the soul is always above the swamp of the materialistic world.

7.8 - Conch

A *shankha* (the shell of a T. pyrum, a species in the gastropod family Turbinellidae) is referred to in the West as a conch shell, or a chank shell. This shell is an important ritual object. It is a ceremonial trumpet through many religious practices, for example during *pooja*. The chank trumpet is sounded during worship at specific points, accompanied by ceremonial bells and singing. As it is an auspicious instrument, of purity and brilliance, it is often played during *Lakshmi pooja* in a temple or at home as one can listen to the sound of the conch by placing it next to the ear. It sounds very similar to Om and is said to drive away negativity. The blowing of the conch or *shankha* needs tremendous power and respiratory capacity. Hence, blowing it daily helps keep the lungs healthy.

7.9 - Bell

The bell is generally made out of brass. A clapper attached to the inside of the bell which makes a high-pitched sound when rung. A brass figure usually adorns the top of the bell handle. Bells are generally hung from the temple dome in front of the *Garbhagriha*. Devotees ring the bell while entering into the sanctum. The sound from the bell echoes Om and welcomes divinity which makes it auspicious. The sound of the bell is said to disengage mind from ongoing thoughts, thus making the mind more receptive. The bell ringing during prayer is said to help in controlling the ever-wandering mind and focusing on the prayer. A mantra is chanted while ringing the bell:

|| *Aagamaardhamtu Devaanaam Gamanaardhamtu Rakshasaam,*
Kuru Ghantaaravam Krutva Devataahvaana Lanchanam ||

This means:

I ring this bell indicating the invocation of divinity, so that virtuous and noble forces enter and the demonic and evil forces depart from within.

7.10 - Symbolism of Deities

As mentioned before and elaborated in Chapter 6, Prayers, the whole universe works on frequencies, and similar frequencies find each other. For this reason, the scripture has its logic and guidance for worshiping God in different forms i.e. Vishnu's conch stands for five elements and eternity, the discus is the symbol of the cosmos, the bow symbolizes power, and the lotus, the cosmos. Shiva's trident represents the three *Gunas* i.e. *Satvik, Rajasik* and *Tamasik*. Shiva is also symbolised for the three lines on his forehead. Similarly, Krishna's flute symbolises divine music. Brahma holds the *Vedas* in his hands, which signifies that he has the supreme control over creative and spiritual knowledge. In the same way, Krishna can be identified by the peacock feather he wears on his head and by a prong-like mark on his forehead. Lakshmi sits on a lotus, implying that she can stay with us in the form of wealth. She also holds a pot with nectar (Amrit Kalash), denoting her capacity to bestow us with great fortune.

Deities are often represented as having a special vehicle for each of them. These are actually our vices, which these deities have the power to overcome. It signifies that if these deities bestow grace on us, we too can ride these vices. Shiva rides the bull, symbolizing brute power. Hence unchecked intensity in anything can be overcome by his grace. The mouse is known for its timidity; Ganesha rides it implying that he has the power to control vices of meekness and timidity. The peacock, the vehicle of Saraswati, denotes that she has control on performing arts. In the same way, Vishnu's serpent denotes human consciousness. Kali rides on a lion, which symbolizes the vices of mercilessness, anger and pride.

7.11 - Temples

A temple is a structure reserved for religious and spiritual rituals such as prayer, effectively used in Indian culture. They are constant companions in the pursuit of truth and ultimate happiness. Though on surface these symbols look trivial and puerile, they have a profound meaning. The word temple is understood to mean *Shivalaya, Mandir, Devalaya, Ambalam, Degul, Kavu, Koil* and a few others.

A *Shivalaya* (where only the *ShivJyotirLinga* is placed) is a symbolic house. The seat and dwelling of the deity is a structure designed to bring human beings and gods together. Inside its *Garbhagriha,* the innermost sanctum, Indian temple contains a murti or the image of the deity or deities in some or other posture. Indian Temples are large and magnificent with a rich history. There is evidence of the use of sacred land as far back as the Bronze Age and later in the Indus Valley Civilization.

7.12 - Brahma

The Brahma Purana describes that there was nothing, but an eternal ocean. From which, a golden egg, called Hiranyagarbha, emerged. The egg broke open and Brahma, who had created himself within it, came into existence (gaining the name Swayambhu). Then, he created the universe, the earth and other things. He also created people to populate and live on his creation. The origins of Brahma are uncertain, in part because several related words such as one for Ultimate Reality.

This is not very far from the scientific theory about the universe we know today, as the referred ocean before this universe may not be the ones we know of our planet earth and the 'egg broke open' referred to as the big bang. As Explained before Brah in Sanskrit means Greater and Atman means soul. Now if we see the Brahma, described in Scriptures, his four faces can be symbolic as four dimensions as x, y, z and time, where we and our world exist.

8 – THE DUTIES

8.1 Five Daily Duties

Deva Yagna (one should worship God daily with a true heart), *Brahma Yagna* (one should study the scriptures and follow the path enunciated therein), *Pitra Yagna* (following the teachings of gurus/ sages/ parents/ elderly people and respecting their seniority, experience and knowledge), *Bhuta Yagna* (helping the needy with care, love and attention) and *Nara Yagna* (we must respect our guests and fellow beings) are the five daily duties were told to be a good human being.

8.2 Five Compulsory Duties

Dharma Karma (following the path of truth and doing righteous), *Tirthyatra* (one should visit holy places and take elders on a pilgrimage), *Utsava* (one should observe and celebrate festivals in the right spirit with family, friends and neighbours and learn the right meanings of rituals associated with the particular festival), *Samskaras* (performing the rituals and following the traditions) and *Sarva Brahma* (the realisation of God) – these are told to be the five essential duties.

8.3 Ten Observances

The ten observances are - *Dhruti* (be patient, firm and stable), *Kshama* (practice forgiveness), *Dama* (be content and controlled), *Asteya* (never steal, conceal or be selfish), *Saucham* (maintain cleanliness, purity and honesty) *Indriya Nigraha* (control your senses and physical desires), *Dhee* (correctly understand the scriptures), *Vidya* (attain and spread the knowledge), *Satya* (always be truthful and stand by the truth) and *Akrodha* (never be angry).

8.4 - Varna Dharma (Professional Duties)

The *Varna* known as the caste system is perhaps one of the most explosive topics, which very sparks controversies. Popular misconceptions are that the Vedic religion encourages division of human beings based on one's birth. As a result, some people were left backward and uneducated while others have used this misconception and misinformation for their personal gains. Much of this misconception can be attributed to the use of the words '*Varna*' and '*Jaati*' interchangeably. A closer analysis will reveal just how wrong these misconceptions are.

In today's time at a Government or private organisation, some people are given higher positions having more authority, salaries and perks due to their required education, skills and experience and some may be doing more hard work, but work on lesser salaries and on smaller positions.

The word caste stands for *jaati*, and is not a word that defines *varna*. It originates in the Portuguese word *casta* which means race, breed, race or lineage. The root word for Varna is '*Vri*' which means occupation. The *varna Dharma* was a division of labour on the basis of their respective work profiles. This division was solely based on the attitude of an individual and his/ her propensity for performing certain duties according to *Gunas* (qualities). There are three *gunas* – *Sattva* (white), *Rajas* (red), and *Tamas* (black).

The issue of *varna Dharma* is highly misunderstood. There are many issues and reasons for the decline of *varna Dharma*. A key point in the *varna Dharma* is its definition and the *gunas* associated with various classes of *varnas*:

Brahmins: priests, scholars and teachers

Kshatriyas: rulers, warriors and administrators

Vaishyas: agriculturalists and merchants

Shudras: labourers and service providers

The *varna* system failed due to many reasons. It even reached the level of untouchability, which turned an ugly asininity on humanity. It was the biggest dividing factor in Indian history. In the beginning, the logic behind this may have been that touching sanitation workers could make sick, as they worked in unhygienic environments. Whatever the reason, Failed *Varna Vibhajan* turned into Castism, Racism and Failed Slavery system has no space in a healthy society. It is for good that as people evolved, this entire system is getting eradicated as we move towards a more equal society. I sincerely hope one day we are entirely free of this system.

One needs to be the change himself first to see the desired change, I have disowned this Failed *Varna Vibhajan* turned into Castism theory long back, and sincerely expect the same from you, the readers of a democratic world. If people start disowning it on their own, it ensures no gains to the opportunists with their agendas and to politicians for their caste based politics for their personal gains and agendas, and will be eradicated in its entirety. In my all professional life, I have kept my late loving father's first name as my surname. In today's era, anybody can be in any profession, including of a priest, teacher or just any profession. Sages like Balmiki of ancient times and then Saints like Ravidas, Kabir, Namdev are many examples. A traditional or caste based surname is not required anywhere anymore; nor makes any significance, and all the people should voluntarily accept this fact, which will help and strengthen our society for the upcoming challenges mankind has in its way ahead.

8.5 - Ashrama Dharma (Family obligations)

As per scriptures, our entire human existence was divided into four stages for better living. The four *purushārthas* are often discussed in the context of four ashramas or stages of life (*Brahmacharya* – student, *Grihastha* – householder, *Vanaprastha* – retirement and *Sannyasa* – renunciation). Scholars have attempted to connect the four stages to the four *purushārthas*. But neither ancient nor medieval texts of India state that any of the first three *ashramas* must devote themselves predominantly to one specific goal of life.

Each of the four canonical *purushārthas* was subjected to a process of study and extensive literary development throughout Indian history. This produced numerous treatises, with a diversity of views, in each category. Some *purushārthas*-focused literature includes

- On *Dharma*

This text discusses *Dharma* from various religious, social, moral and ethical perspective. Each of the six major schools has their own literature on *Dharma*. Examples include *Dharmasutras* (particularly by Gautama, Apastamba, Baudhayana and Vāsistha) and *Dharmasastras* (particularly Manusmriti, Yājñavalkya Smriti, Nāradasmriti and Vishusmriti). At a personal *Dharma* level, this includes many chapters of *Yogasutras*.

- On *Artha*

Artha-related texts discuss from individual and social standpoints, and as a compendium of economic policies, politics and laws. For example, the *Arthashastra* of Kautilya, the *Kamandakiya Nitisara, Brihaspati Sutra*, and *Sukra Niti*. Olivelle states that most *Artha*-related treatises from ancient India have been lost.

- On *Kama*

These discuss arts, emotions, love, erotica, relationships and other sciences in the pursuit of pleasure. The *Kamasutra* of *Vātsyāyana* is one of the most well-known texts. Others texts include the *Ratirahasya, Jayamangala, Smaradipika, Ratimanjari, Ratiratnapradipika, Ananga Ranga* among others.

- On *Moksha*

These texts debate the nature and process of liberation, freedom and spiritual release. Major treatises on the pursuit of *moksha* include the *Upanishads, Vivekachudamani*, Bhagavad Gita, and the *shastras* on yoga.

9 – THE CUSTOMS

9.1 The Meaning of Samskara

There is no right English word for *samskara*. *Samskara* are rites of passage in a human being's life described in ancient Sanskrit texts, as well as a concept in the Karma studies of Indian philosophies. The word literally means 'putting together, getting ready', or 'a sacred or sanctifying ceremony' in ancient Sanskrit.

Customs are not merely rituals. In fact, they are a testimony to the importance accorded to spiritualism and knowledge in our culture. They are also backed by the scriptures. They are user friendly and programmed to inculcate good values in people, for the greater quest to the truth as well. It was asked to follow a pious path acceptable to God, hence we imbibe good qualities in our progeny. These intentional efforts are called *Samskaras*. In every stage of human life, certain rituals serve as a constant reminder of the importance and the goal of human life. In the context of *Karma* , *Samskaras* are dispositions, character or behavioural traits, that exist by default from birth, or are ingrained and perfected by a person over their lifetime. They exist as imprints on the subconscious according to various scriptures. These perfected or default imprints of Karma influence that person's nature, responses and states of mind. The child, like a sapling, is nurtured with utmost devotion and attention. From the very beginning good *Samskaras* are taught so that the child can grow into a perfect gentleman who will understand himself and the world around him with ease. These are the *samskaras* that, if properly followed, will take one in the right direction towards eternal truth: *Garbhadhana* (consummation of the marriage), *Pumsavana* (celebrating the fetus), *Simantonnayana* (parting of the pregnant woman's hair in the 8th month), *Jatakarman* (the birth), *Namakarana* (naming the child), *Annaprashana* (the baby's first feeding of solid food), *Choulam* (baby's first haircut, tonsure), and *Upanayana* (thread ceremony, *janeyu* and entry into school rite); *Vivaha* (marriage); *Sanyaas* (mediation) and *Anthyesthi* (cremation after death)

To obtain union with the *Brahman*, one must also possess the eight virtues (compassion, patience, non-envy, purity of thought, speech and body, inner calm and peace, positive attitude, generosity, and lack of possessiveness).

9.2 - *The Cow*

The cow is considered sacrosanct. There are many medicines, antibiotics, pain killers, and skin care cosmetics made by cow formulations, from their milk to excreta. According to research, cow milk has great chemical properties and contains antibodies that are good for human beings. In agrarian society, which plays the greatest role in human evolution, the utility of a cow is unlimited, so it has its own place of respect for its worth.

9.3 - *The Lights (Deepa)*

Light symbolizes knowledge. God is knowledge and the creator of knowledge. Knowledge is everlasting wealth and removes ignorance just like light removes darkness. Hence the lighting of a lamp symbolises bowing down to knowledge as the greatest of all powers. Oil and ghee symbolise our *vaasanas* or vices, and the wick, the ego. When lit by spiritual knowledge, the *vaasanas* slowly get exhausted and the ego perishes. The flame of a lamp always burns upwards against gravity, symbolising that we should acquire knowledge to take us towards higher ideals.

9.4 - The Coconut

A coconut looks like the head of a human being. It grows on sandy earth and uses salty water, but gives us sweet water enriched with healthy nutrients in it. Breaking of the coconut symbolizes the breaking of our pride. The water within represents our weakness. When offered along with white kernel that represents the mind, it is believed that the touch of God purifies our minds. A mind thus purified is passed on as *prasada*, a holy gift, and the same is reverently accepted by all.

9.5 - Touching the feet

Touching the feet of elders is a sign of respect for the experience, maturity, nobility and divinity that our elders personify. It is our recognition of their selfless love and the sacrifices that they have made for our welfare through their lives. It is a way of humbly acknowledging the greatness of another. Moreover, when elders give blessings, a positive energy is passed on to us.

9.6-Namaste

Saying *namaskara* or *namaste* in Sanskrit means 'I bow to you, and the God within you'. *Namaha* is literally interpreted as

'*Na Ma*' meaning not mine. It has the spiritual connotation of reducing one's ego in the presence of the other. *Anjali* is the Sanskrit word for 'divine offering' or a gesture of reverence, and is derived from *anj*, meaning 'to honour or celebrate'.

Mudra means 'seal' or 'sign'. The meaning of the phrase is thus a seal for salutation. *Anjali mudra* is performed by pressing the palms of the hands together. The fingers are pressed together with fingertips pointing up, with the palms touching evenly.

In the most common form of anjali mudra, the hands are held at the heart chakra, with the thumbs resting lightly against the sternum. The gesture may also be performed at the *ajna* or brow *chakra* with the tips of the thumbs resting against the third eye or at the crown *chakra* between the eyebrows. In certain yoga postures, the hands are placed in *anjali mudra* position to one side of the body or behind the back.

Anjali mudra is normally accompanied by a slight bowing of the head.

The gesture is also known as *hridayanjali mudra* meaning 'reverence to the heart' seal (from *hriday*, meaning heart) and *atmanjali mudra* meaning 'reverence to the self' seal (from *atman*, meaning self). The symbolic *anjali mudra* has the same meaning as the Sanskrit greeting *namaste* and can be performed while saying *namaste* or *pranam*, and can also be used instead of saying the actual words.

The gesture is used for both greetings and farewells, but carries deeper significance than a simple hello or goodbye. The folding of the palms is said to provide a connection between the right and left hemispheres of the brain and represents unification. This yoking is symbolic of the practitioner's connection with the divine in all things. Hence, *anjali mudra* honours both the self and the other.

In yoga, *anjali mudra* is a centering pose, which according to practitioners, helps to alleviate mental stress and anxiety and therefore is used to assist the practitioner in achieving focus and entering a meditative state. The physical execution of the pose helps to promote flexibility in the hands, wrists, fingers and arms.

9.7 - The Tilak

The *tilak* covers the spot between the eyebrows, which is the seat of memory and thinking. It is where the pituitary gland is located, and is called the *ajna chakra*. The three *Kundalini Nadies* - *Ida, Pingala* and *Sushumna* converge here. The *tilak* is applied with the prayer: may the presence of God pervade within me all the time and encourage me to be truthful in all that I do. The presence of this mark reminds us of our resolve throughout the day. The *tilak* is thus a blessing and a protection against wrong vices and tendencies. The entire body emanates energy in the form of electromagnetic waves – the forehead and the subtle spot between the eyebrows especially so. The *tilak* made from sandalwood paste cools the forehead, protects it, and prevents energy loss. My personal experience of applying a sandalwood powder *tilak* on my forehead has always been a great way to maintain calmness and peace.

9.8 - The Offering (Bhog)

Everything in the world is given to us by the grace of the God, who is everything including his grace itself. Hence before consuming, it is offered to God first as a symbolic gesture. Thereafter it is his gift to us, graced by his divine touch. Knowing this, our entire attitude to food and the act of eating it changes.

9.9 - Sprinkling of water

This is a symbolic act of purification. Five morsels of food are placed on the side of the plate acknowledging the debt owed to the divine energies (*Devta Runa*) for their benign grace and protection; our ancestors (*Pitru Runa*) for giving us their lineages; the sages (*Rishi Runa*) for giving us this profound knowledge and understanding of God, our fellow beings (*Manushya Runa*), and other living beings serving us selflessly.

9.10 - Bhasma

The application of *Bhasma* (ash) signifies the destruction of evil and remembrance of the divine. *Bha* means to destroy and *sma* implies to remember. *Bhasma* is also called *vibhuti* (glory) as it gives glory to one who applies it, and raksha (*protection*) as it protects the wearer from ill health and evil, by purifying him or her.

9.11 - Homa

Homa is again a symbolic oblation to fire with sacred chants signifying the surrender of ego and negative desires in the flames of knowledge. The consequent ash signifies the purity of the mind, which results from such actions. Also, the fire of knowledge burns the oblation and wood, signifying the burning of ignorance and lethargy respectively. The ash we apply indicates that we should burn false identification with the body and become free from the limitations of birth and death.

9.12 - Apology

Touching another person with the feet is akin to disrespecting the divinity within him or her. The same applies to touching a book, currency, idols, or other pious things with the foot. One must immediately apologise if something of the kind happens.

9.13 - Tulsi

The tulsi plant has the potency to cure and is used in many ayurvedic preparation. Women worship tulsi early in the morning, mainly to inhale fresh air. It is known that tulsi releases the maximum amount of oxygen. Tulsi worship is a breathing exercise and the same applies with peepal (Ficus Religiosa tree), as that exhales oxygen even when asleep.

9.14 - Aarti

Burning camphor signifies that our false ego and negative tendencies too shall burn. When the camphor burns, it spreads a pleasant fragrance. In the same way we should also sacrifice our material possessions to spread love and peace. During *aarti*, your eyes remain shut to introspect within; this signifies that each of us is a temple of God. At the end of the *aarti*, the hands are placed above the flame and then touched to the eyes and over the top of the head, signifying the divine light brightening our vision and thoughts.

9.15 - Kalava or Kautuka

The *pratisara* or *kautuka* is a red-yellow coloured protection thread. It is mentioned in a ritual thread context in the Vedic text *Atharvaveda Samhita* 2.11. Either a priest or the oldest family member of the family ties the protection thread on the wrist of a devotee, patron or loved one. The thread is traditionally worn on the right wrist by males and on the left by the females. In regional Vaishnavi tradition, the thread symbolizes the divinity of the deities as Vishnu for men and Lakshmi for women.

9.16 - Janeyu Upanayana

Janeyu Upanaya in Sanskrit literally means 'leading to or getting near the eye (or eyesight / vision/ third eye/ pineal gland)'. It is an important and widely discussed *samsara* in ancient Sanskrit texts. The rite of passage symbolises the leading or drawing of a child towards the self in a school by a teacher. This initiates the second birth of the young mind and spirit. *Upanayana* is the rite of passage for the start of a person's formal education of writing, counting, reading, studying *Vedangas*, arts and other skills. The *Upanayana* rite of passage was also important to the teacher, as the student would thereon begin to live in the gurukul (school). The sacred thread or *janeu* is called by yet another name, *yajnopaveetam,* suggesting that the individual wearing it possesses an intense desire to act according to the highest good (*Yajna*). *Janeu* is made of three strands of cotton thread known as *vritas*. Every thread has three cords braiding around each other and a knot. *Janeu* is associated with a lofty kind of symbolism that guides the boy to live an ideal life. It is worn on the left shoulder, symbolising the shouldering of the burdens of life with tolerance. It runs over the heart, indicating a life of faith and determination. It runs across the back, symbolising commitment. Thus, symbolically janeu enjoins a man to a life of commitment, courage, determination, faith, patience and tolerance.

9.17 - The Shikha

This is a knotted tuft of hair on top of the head, also known as *choti*. Like the *tilak* and *janeyu*, a *shikha* is also a prominent symbol of *Santana Dharma* followers. The few tied hairs at the top back of the head applies very little pressure on particular point of the head, that helps improve concentration and self control, and improve memory. There is a super sensitive spot on the head known as *Adhipati Marma*, where there is a nexus of all nerves. The *shikha* protects this spot. Below it, in the brain, occurs the *Brahmarandhra,* where the *sushumnā* (nerve) connects from the lower part of the body. In yoga, *Brahmarandhra* is the highest chakra, the seventh chakra with the thousand-petalled lotus. It is the centre of wisdom. The knotted shikha helps boost this centre and conserve its subtle energy known as *Ojas*.

9.18 - Om Shanti

Frequently chanting *shanti, shanti, shanti* brings peace. It is believed that *Trivaran Stayam,* that which is said thrice, come true. It is repeated thrice to emphasise our intense desire for peace.

10 – THE FESTIVALS

10.1 - Principles of Vedic Times

Before understanding the logic behind the festivals, let us understand the profound ancient principles at the time of Vedic studies. *Kaala,* or time, is accorded much importance in our lives. According to the *Vedas,* time is spherical and hence called *Kaalachakra,* the wheel of time. It has no beginning and no end. It is amazing that much before the advent of the clock our ancestors had devised a method to measure time to such micro levels.

1 Parmanu = 1/379675 of a second

2 Paramanu = 2 Anu

3 Trisarenu = 1 Trithi

100 Trithi = 1 Vedha

3 Vedha = 1 Lava

3 Lava = 1 Nimisha

3 Nimisha= 1 Kshana

5 Kshana= 1Kashta

15 Kashta= 1 Laghu

15 Laghu = 1Ghatika (24 minutes)

2 Ghatika = 1 Muhurta

3.75 Muhurta = 1 Prahara

$$8 \text{ Prahara} = 1 \text{ day}$$

$$15 \text{ days} = 1 \text{ Paksha}$$

$$2 \text{ Paksha} = 1 \text{ Maasa (month)}$$

$$2 \text{ Maasa} = 1 \text{ Ritu}$$

$$6 \text{ Ritu} = 1 \text{ Varsha}$$

Festivals also hold a special significance. Apart from their *Dharmic* importance, they have philosophical and logical significance. All Indian festivals are celebrated according to planetary movements and have scientific significances. India has been an agrarian society owing to its vast fertile planes, rivers and mineral-rich soil and minerals. A large number of harvest and sowing festivals are observed across India. After the sowing periods, farmers have ample time at their disposal, and many religious activities are organised. This is in order to inculcate religious feelings and further community living. After the harvest, festivals are celebrated where fireworks are used to kill the insects and germs that threaten the crops after the rain.

In Vedic tradition, there are many festivals observed throughout the year. There are a few major festivals and numerous minor ones, as well as those celebrated on a local or regional basis. There are different festivals to celebrate the various incarnations of deities, as well as those that honour the seasons, harvest cycles, relationships, and certain other principles of Vedic culture. Some of the major festivals are covered in the next few sections.

10.2 - *Makara Sankranti*

Vedic scriptures consider the sun the king of the planets. *Makara Sankranti* is the celebration of the sun's journey to the Northern hemisphere. *Makar* means Capricorn and *Sankranti* means transition. You could say there is one *sankranti* every month, when the sun moves from one sign of the zodiac to the next. However, the most important ones are the *Mesh* (Aries) *Sankranti* and the Makar (Capricorn) *Sankranti*. So the transition of the sun from Sagittarius to Capricorn around the winter solstice in the Northern hemisphere is known as Makar *Sankranti*. This is when the sun moves from the *Dakshinayana* (Southern) route to the *Uttarayana* (Northern) route. The *Uttarayana* route begins on January 14 and lasts until July 14.

This is an auspicious time. The Vedic scriptures say that those who die in this period attain better chances of *moksha* or liberation from *samsara*, the cycle of birth and death. It is also a time when taking a dip in a holy *sangam* (a confluence of holy rivers) is especially befitting. In line with astrological calculations, the *Kumbha Mela* is conducted once in 12 years at Prayagraj (Allahabad) beginning on *Makara Sankranti*. This is the day when the sun and moon enter Capricorn and Jupiter enters Aries. The astrological configuration on *Makara Sankranti* is called 'Maha-snana-yoga', the day for taking a special holy bath, and a highly auspicious time that allows the soul easy access to the celestial worlds.

Makar Sankranti falls on January 14 and in leap years on January 15. It is the only festival that is based on the solar calendar instead of the lunar calendar.

Makar *Sankranti* is celebrated as a harvest festival. It is a way of expressing thanks to Mother Earth, and for spreading goodwill, peace and prosperity. It is celebrated at the time when winter starts to recede, giving way to spring and summer. People also give each other presents and traditional sweets made with sesame, and throw feasts that includes a dish made of green gram, rice and jaggery.

10.3 - Vasant Panchami

Vasant Panchami is also called the festival of kites. It is celebrated towards the end of winter, around the months of January-February. *Vasant Panchami* is celebrated in the Northern parts of India. The weather changes from harsh winter to soft spring, called *'vasant'*. *Vasant* is the time when the mustard fields abound with yellow flowers that seem to usher in spring. People welcome the change of season and celebrate the day by wearing yellow clothes, holding feasts and organizing kite flying. *Vasant Panchami pooja* is devoted to Saraswati, the Goddess of learning and knowledge.

10.4 - Maha Shivaratri

Shivaratri is celebrated sometime between February-March. It is believed that Parvati (the symbol of energy), the wife of Shiva (the symbol of matter), prayed, meditated and fasted on this day for the well-being of Shiva and hoped to ward off any evil that could fall upon him. Even though both men and women celebrate *Shivaratri*, it is an especially auspicious day for women. Married women pray for the well-being of their husbands and sons, while unmarried women pray for a husband like Shiva, the ideal husband.

On *Shivaratri*, devotees wake up at sunrise and bathe in holy water (preferably in the River Ganga) and wear new clothes. On the day of the festival, people fast and spend the day focused on Shiva, meditating and chanting '*Om Namaha Shivaya*'. This holds the mind in single-pointed focus throughout the day. Then they go to temples carrying holy water to bathe and worship the *ShivaLingam*, and to decorate it with flowers and garlands. This bathing of the *ShivaLingam* symbolizes the cleansing of one's soul. It is customary to spend the entire night singing praises to Lord Shiva.

Shivaratri is observed in the pattern of preparation, purification, realization, and then celebration. Then at the stroke of midnight, Shiva is said to manifest as the inner light of purified consciousness. This climax at night represents overcoming the darkness of ignorance and reaching a state of purified spiritual knowledge. In this way we conquer the influence of the mind and the senses, and enter the state of steady awareness, where there is spiritual awakening. If one can follow this process, then he or she can experience the real meaning of *Shivaratri*.

10.5 - Holi

The word *ho-li*, literally means 'let-go' in English. It is a major festival and celebrates the onset of spring, along with a rich harvest and fertility in the land. It is celebrated on the day after the full moon in early March across India. It is known around the world as the festival of colours, as people throw bright coloured powders and water over each other to celebrate the advent of spring. Vibrant processions accompanied by folk music and dancing are also a characteristic of *Holi* celebrations, along with the distribution of sweets and new clothes. *Holi* is a very popular festival amongst the youth. Huge bonfires are lit on the eve of *Holi*. Religiously, it is a reminder of the death of Holika, the aunt of Prahlada, by burning her in a bonfire, the symbolic gesture of destroying all negativity within and accepting everyone with love and rejoicing.

10.6 - Rama Navami

This festival marks the birth of Lord Rama, an incarnation of Lord Vishnu. It is celebrated around March-April. Lord Rama, who became the king of Ayodhya, was known for his exemplary qualities. He was popular, brave, kind, just, intelligent, patient, loving, obedient and dutiful. Lord Rama is always worshiped with his wife Sita, brother Lakshmana and devotee Hanuman. The worship of Lord Rama is accompanied by the worship of the Sun god, since Rama was considered to have descended from the sun, in the Solar dynasty. *Rama Navami* celebrations include reading the Ramayana and staging plays about the *Rama Lila,* the life of *Lord Rama.*

10.7 - Ugadi and Vishu

These are two festivals mark the beginning of the New Year in different communities of South India. Ugadi is celebrated in March-April. Vishu is celebrated in mid-April. The word ugadi means the day of the inauguration of the Yuga or age. Vishu is celebrated in a big way in Kerala. Families wake up in the morning and feast their eyes on things like a picture of God, grains, flowers, fruit and gold. It is believed seeing these first thing in the morning of the New Year will bring them prosperity and wealth throughout the year.

10.8 - Hanuman Jayanti

This celebrates the birth of Hanuman, the most ardent worshipper of Lord Rama. His birthday falls on *Chaitra Shukla Purnima*—the March-April full moon day. On this holy day, people fast and worship Lord Hanuman, read the *Hanuman Chalisa* and spend the day chanting *Ram-Nam*. Celebrations are marked by special *poojas* for Hanuman.

Hanuman testified to the power of *Ram-Nam* through his life. He was an ideal selfless worker, a true devotee who worked without personal desires, and an exceptional Brahmachari or celibate. He lived only to serve Lord Rama with pure love and devotion. He was humble, brave and wise. He possessed all the divine virtues - devotion, knowledge, a spirit of selfless service, power of celibacy, and desirelessness. He never boasted of his bravery and intelligence.

10.9 - Guru Purnima

Guru Purnima celebrates the might of one's teacher or guru with respect and reverence. Also known as *Vyasa Purnima*, the festival is celebrated in July-August on the full moon day. It is believed that the great scholar and composer of Vedic literature, Vyasadeva, who lived in the *Dvapara Yuga*, was born on this day. Legend also has it that this was the day he completed the codification of the four *Vedas*.

10.10- Onam

As per scriptures, this festival marks the day on which the devotee of Lord Vishnu, Emperor Maha Bali, the grandson of Prahlada, received benediction and liberation with the blessings of the Lord, who had assumed the form of Vamanadeva, his dwarf incarnation. Onam is celebrated in August-September, majorly in Kerala. Onam is a ten-day festival marked by creating beautiful floral patterns in front of their houses, leading *poojas* for Lord Vishnu, feasting with the whole family and watching boat races.

10.11- Raksha Bandan

This festival celebrates the love of a sister and brother. On this day, sisters tie a *rakhi,* a colourful bracelet made of silk thread, on the wrist of their brothers to protect them against evil influences. It is celebrated in July-August.

10.12- Krishna Janmashtami

Simply known as *Janmashtami,* this is the celebration of the birth of Lord Krishna. It is observed on the eighth day of the dark fortnight in August-September. Temples and homes are beautifully decorated and lit. Notable among these decorations are the cribs and other decorations that depict stories of Lord Krishna's childhood. In the evening, *bhajans* (devotional songs) are sung till midnight, the auspicious moment when Lord Krishna was born. On the day of the festival, people fast and spend the day focused on Lord Krishna, meditating and chanting the *Hare Krishna* mantra and other prayers or songs devoted to him. Often there are also plays and enactments organised, of the birth and life of the Lord. Thus, offering their obeisance, focusing their minds on Lord Krishna, devotees hold themselves in such single-pointed focus throughout the day. Then at the stroke of midnight, Lord Krishna takes birth, which is celebrated by a midnight aarti ceremony. Flowers are showered on the deity of Lord Krishna, or the deities are dressed in new outfits.

In this way, after a full day of purification, we realize our own connection with the Lord, who then manifests as the supreme object of our purified consciousness. Thus, this climax at night represents our overcoming the darkness of ignorance and reaching a state of purified spiritual knowledge and perception. If one can follow this process, then he or she can experience the real meaning of *Krishna Janmashtami*. Then *prasada* is distributed to everyone.

10.13 - Ganesh Chaturthi

This celebrates the birth of Lord Ganesh, also known by his child form Vinayaka in southern India. He is the deity of wisdom, prosperity and good luck, and the remover of obstacles. Ganesh Chaturthi is celebrated on the fourth day of the lunar month that falls in August-September, for a period of two days, or in some cases ten days. Clay figures of the elephant-headed Ganesh are made and, after being worshiped, are submerged in water. Ganesh Chaturthi is very popular in the state of Maharasthra.

10.14 - Navaratri

Navaratri or the nine sacred nights dedicated to the Mother Goddess is celebrated in the month of October-November. *Navaratri* includes the Saraswati Pooja and the Durga Pooja festivals. 'Nava' means nine and 'ratri' means night. So Navaratri literally means nine nights. It is during these nine nights of festivity that the goddess is worshiped in her different forms - Durga, Lakshmi and Saraswati. Durga is worshiped during the first three nights of the festival because of her destructive nature. She destroys the *anarthas* or unwanted barriers that hold us back from our true spiritual potential, and removes negativity from the mind, which is the meaning of *durgati harini*. Without removing these obstacles, spiritual unfolding cannot take place.

The next step is the positive process of adding the qualities we need. So the goddess of wealth Lakshmi is worshiped over the next three nights. She is the goddess of love, goodness, compassion, forgiveness, cooperation, nonviolence, devotion, purity, and the like. Virtue is the true wealth, which is given by Lakshmi. This is not merely riches and possessions, but spiritual wealth that can propel us toward the spiritual goal. These positive qualities replace the bad ones that were removed by Durga.

At this point, the seeker becomes fit for philosophical study and contemplation that is required. Then Saraswati the goddess of knowledge is worshiped during the remaining three nights. Saraswati gives one the intelligence, knowledge and wisdom by which spiritual realisation is possible. She represents the highest knowledge of the self. She plays her well-tuned *veena* of knowledge and insight, which can then tune our mind and intellect to work in harmony with the world and the purpose of our existence. Then our spiritual practice, study, and meditation become effective in rising above the influence of our mind and senses. Then we can perceive our real identity of being spiritual entities and parts of the spiritual dimension, free from illusion.

After having removed our impurities, gained the proper virtues, and then acquired the knowledge of the self, we come to the last day. This day is called *Vijayadashami*, or the day of victory over our minds and the lower dimension after having worshiped the goddess in her three forms. The celebrations of *Navaratri* are held at night because it represents our overcoming the ignorance of the mode of darkness, the night of *tamoguna*.

Additionally, *Navaratri* commemorates the day on which the combined powers of the three Goddesses of Durga or Maha Kali, Maha Lakshmi and Maha Saraswati put an end to the evil forces represented by the buffalo-headed demon Mahishasura. The ninth day is also the day of the *Ayudha Pooja* in the South, the worship of whatever instruments one may use in one's livelihood. On the preceeding evening, it is traditional to place these instruments on an altar to the Divine. If one can make a conscious effort to see the Divine in the tools and objects one uses each day, it will help one to see one's work as an offering to God. It will also help one to maintain constant remembrance of the Divine. Children traditionally place their books and writing implements on the altar. Throughout the ninth day, people make an effort to see their work or studies as imbued with the divine presence. The tenth day is called *Vijayadashami*. Devotees perform a *pooja* to the Goddess Saraswati to invoke her blessings on books, writing implements, musical instruments and tools of their trade. After the *poojas*, little children are initiated into the learning process.

10.15 - Dusshera

Also known as *Vijayadashami, Dusshera* is celebrated on the tenth day of *Navaratri*. This signifies the victory of Lord Rama over the demon Ravana, which is marked with the burning of the effigy of Ravana. The effigy is often burned in a huge bonfire or with a flaming arrow, which also represents the destruction of the false ego. Thus, it is a festival which shows the process by which humanity can reach the perception of God. It incorporates the means by which one can purify themselves of the ten sins, meaning the sins committed by the ten active senses. It is the process of purification, so that one can become free of the temporary world and materialism, paving the way to a transcendental experience.

What this shows is that all aspects of the Vedic process, whether we are familiar with them or not, are ultimately meant to be a vehicle by which we can transcend the mind, senses, and the temporary material world and enter into the supreme reality wherein we can re-establish our lost relationship with God.

10.16 - Karva Chauth

This is a fast undertaken by married women who offer prayers seeking welfare, prosperity and longevity of life for their husbands. *Karva Chauth* is celebrated before *Deepavali,* sometime in October or November. It is the most important fast observed by the women of North India. A woman keeps such a fast for the welfare of her husband, who becomes her protector after she leaves her parents' home. This is a tough fast to observe as is starts before sunrise and ends after worshiping the moon, which usually rises late in the evening.

10.17 – Deepavali

Deepavali, more popularly *Diwali*, is the festival of lights. It symbolizes the victory of righteousness and the lifting from spiritual darkness. The word *Deepavali* literally means rows of clay lamps. It is celebrated on the new moon day of the dark fortnight during October-November. It is also associated with the return to Ayodhya of Lord Rama, his wife Sita and his brother Lakshmana after their fourteen-year sojourn in the forests. The day also marks the coronation of Lord Rama.

The meanings of *Diwali*, its symbols and rituals, and the reasons for its celebration are innumerable. *Diwali* signifies the triumph of good over evil, of righteousness over treachery, of truth over falsehood, and of light over darkness. *Diwali* is celebrated every year with family, fun, festivity and fireworks.

Diwali also marks the beginning of the New Year. For some, the day of *Diwali* itself is the first day of the New Year, and for others it is the day just before the New Year begins. Either way, it is a time of cleaning out homes and offices, letting the light of *Diwali* enter all the corners of our lives. It is a day to start afresh.

Children love the fireworks associated with *Diwali*. A lot of sweets are distributed to friends and relatives. Homes are often lit with rows and rows of little clay lamps called diyas that light up the dark new moon night. Businesses begin their new book keeping on *Diwali*. The trading community celebrates the thirteenth day of the month of Kartika (Oct.-Nov) as Dhanteras or Dhantrayodashi, the first day of a five-day festival. The word Dhan means wealth, and the day is of great importance for the rich mercantile community of Western India. The day ends with a Lakshmi pooja at home. Some temples also conduct large Lakshmi Pooja celebrations.

This is the third, and perhaps most important, aspect of *Diwali*: the worship of Maha Lakshmi. Maha Lakshmi is the goddess of wealth and prosperity, bestowing these abundantly upon her devotees. On *Diwali* we pray to her for prosperity; we ask her to lavish us with her blessings.

The following day is celebrated as Narka-Chaturdashi or Choti *Diwali*. As per the scriptures, Lord Krishna and his wife Satyabhama are said to have returned home victorious after killing demon Narakasura early in the morning on this day. The Lord was massaged with scented oils and was given a good bath to wash away the filth from his body. Since then, on this day, the custom of taking an oil bath with fragrant uptan before sunrise has become a traditional practice in Maharashtra and South India.

The day of *Diwali* is devoted entirely to the propitiation of Goddess Lakshmi, that comes with burning lamps, firecrackers, and card games. On the dark night of Amavasya, businessmen perform *Chopda Poojan* and open new account books.

The day following *Diwali* is the day of *Govardhana Pooja*. The day is also observed as *Annakoot* in temples of Mathura and Nathdwara. According to the Vishnu Purana, years ago the people of Gokul used to celebrate a festival in honour of Lord Indira and worship him after the end of every monsoon season. However, one year the young Lord Krishna prevented them from offering prayers to Indira and convinced the people to offer the *pooja* to Govardhan Hill, since it was an incarnation of the supreme. This enraged Lord Indira, who in turn sent a huge flood to submerge Gokul. But Lord Krishna saved Gokul and all the residents by holding Govardhan Hill up like an umbrella.

The fifth and final day of the *Diwali* festival is known as *Bhaiya-Duj* or *Bhav-Bij*. According to the legend, Lord Yama, the God of death, visited his sister Yami on this day. She applied the auspicious *tilak* on his forehead, garlanded him and served him delicious sweets. In return, Yama gave her a special gift as a token of his love and pronounced that anyone who received *tilak* from his sister would never be defeated. And so to this day, brothers never fail to visit their sisters on the final day of *Diwali*.

With lights everywhere, *Diwali* symbolises the dispelling of darkness, ignorance and evil, and a new hope for the future and irrespective of the region, unites the nation in the celebration of prosperity and joy.

10.18 - Gita Jayanti

This is the celebration of when Lord Krishna spoke the illustrious Bhagavad Gita to Arjuna on the battlefield of Kurukshetra, north of New Delhi. This usually takes place in the early part of December.

The Vedic festivals are performed in these phases of preparation, purification, realization, and then celebration. It represents one's progress toward the real goal of life. First, the mind must be purified of all unwanted thoughts and habits.

Then it must become focused on one's concentration on the Supreme. As the knowledge of our spirituality, the self and our connection with the supreme being is revealed, it is followed by realization. When such realization has been reached and the ego destroyed, then there is celebration. Living life on the basis of spiritual realization means that life is a constant joy and celebration.

"Around one trillion of species are currently living on the earth. This is less than one percent of the total species that ever lived on this planet in its four-billion-years history of life on it.

During the process of evolution, only very few could survive by being responsive, adaptive and by developing their skills needed to be compatible with the only constant factor as 'Change', not just their sheer strength or intelligence.

This book of mine will be remembered saying that the same principle applies on religions too, specially for the ones, which run on fixed rules with dogmas, having no scope for their up-gradation or updates."